Lecture Notes
in Control and Information Sciences   233

Editor: M. Thoma

Springer-Verlag London Ltd.

Pasquale Chiacchio and Stefano Chiaverini (Eds)

Complex Robotic Systems

Springer

ISBN 978-3-540-76265-2

British Library Cataloguing in Publication Data
Complex robotic systems. - (Lecture notes in control and
   information sciences ; 233)
   1.Robotics
   I.Chiacchio, Pasquale   II.Chiaverini, Stefano
   629.8'92
   ISBN 978-3-540-76265-2   ISBN 978-3-540-40904-5 (eBook)
   DOI 10.1007/978-3-540-40904-5
Library of Congress Cataloging-in-Publication Data
A catalog record for this book is available from the Library of Congress

Typesetting: Camera ready by editors

69/3830-543210 Printed on acid-free paper

No' si volta chi a stella è fisso.
Leonardo da Vinci

Preface

The challenges that mankind must face in this era of astonishing progress in technology calls for the development of a common and up-to-date worldwide knowledge base. When working at this book our intention was to realize a small contribution to the achievement of this goal within the field of Robotics.

Robotic systems have proven themselves to be of increasing importance and are widely adopted to substitute for humans in repetitive and/or hazardous tasks. Their diffusion has outgrown the limits of industrial applications in manufacturing systems to cover all the aspects of exploration and servicing in hostile environments such as undersea, outer space, battlefields, and nuclear plants.

Complex robotic systems, i.e. robotic systems with a complex structure and architecture, are gaining increasing attention from both the academic community and industrial users. The modeling and control problems for these systems cannot be regarded as simple extensions of those for traditional single manipulators since additional complexity arises: to accomplish typical tasks there is the need to ensure coordinated motion of the whole system together with management of interaction between each component of the system.

This book focuses on two examples of complex robotic systems; namely, cooperating manipulators and multi-fingered hands.

In April 1997 we organized a Tutorial Session on these topics at the IEEE International Conference on Robotics and Automation held in Albuquerque, NM, collecting contributions from distinguished scientists throughout the world. The collected material was of high quality and up-to-date, thus we thought it could be of interest to a wider audience. Therefore, we asked all the contributors to further extend their manuscripts; all of them agreed and the result of this joint effort is this book.

Although the book is the outcome of a joint project, the individual contributions are attributed as detailed in the following. We feel the need to thank our colleagues for their motivation during the project.

In Chapter 1, Masaru Uchiyama gives a general perspective of the state of the art of multi-arm robot systems. After outlining the historical evolution of studies in this area, he gives the fundamentals of kinematics, statics and dynamics of such systems.

Chapter 2 has been written by John T. Wen and Lee S. Wilfinger. They extend the manipulability concept commonly used for serial manipulators to general constrained rigid multibody systems. The concepts of unstable grasp and manipulable grasp are also introduced.

In Chapter 3 we present the kinematic control approach for a dual-arm system. An effective formulation is presented which fully characterizes a coordinated motion task, and a closed-loop algorithm for the inverse kinematics problem is developed. A joint-space control scheme based on kineto-static filtering of the joint errors is devised and analyzed.

Michael A. Unseren in Chapter 4 reviews a method for dynamic load distribution, dynamic modeling, and explicit internal force control when two serial link manipulators mutually lift and transport a rigid object. A control architecture is also suggested which explicitly decouples the two set of equations comprising the model.

Ian D. Walker devotes Chapter 5 to a survey of design, analysis, and control of artificial multi-fingered hands and corresponding research in the area of machine dexterity. An extensive bibliography is also provided.

In Chapter 6 Friedrich Pfeiffer presents optimal coordination and control of multi-fingered hands for grasping and regrasping. The method is applied to an experimental setup consisting of a hand with hydraulically driven fingers which ensure good force control.

The book is addressed to graduate students as well as to researchers in the field. We hope they will find it useful and fruitful.

Napoli, Italy, September 1997

Pasquale Chiacchio, Stefano Chiaverini

Contributors, in chapters' order, are: *Masaru Uchiyama*, Tohoku University, Japan; *John T. Wen* and *Lee S. Wilfinger*, Rensselaer Polytechnic Institute, U.S.A.; *Pasquale Chiacchio* and *Stefano Chiaverini*, Università di Napoli Federico II, Italy; *Michael A. Unseren*, Oak Ridge National Laboratory, U.S.A.; *Ian D. Walker*, Clemson University, U.S.A.; *Friedrich Pfeiffer*, Technische Universität München, Germany.

Contents

Chapter 1

Multi-arm robot systems: A survey

This chapter presents a general perspective of the state of the art of multi-arm robot systems which consists of multiple arms cooperating together on an object. It presents first a historical perspective and, then, gives fundamentals of the kinematics, statics, and dynamics of such systems. Definition of task vectors highlights the contents and gives a basis on which cooperative control schemes such as hybrid position/force control, load sharing control, etc. are discussed systematically. Practical implementation of the control schemes is also discussed. Implementation of hybrid position/force control without using any force/torque sensors but with exploiting motor currents is presented. Friction compensation techniques are crucial for the implementation. Lastly, the chapter presents a couple of advanced topics such as cooperative control of multi-flexible-arm robots and robust holding with slip detection.

1.1 Introduction

It was not late after the emergence of robotics technologies that multi-arm robot systems began to be interested in by some of robotics researchers. In the early 1970's, they had already started research on this topic. The reason was apparent, that is, due to many limitations in applications of the single-arm robot; the single-arm robot can carry only small objects that can be grasped by its end-effector, needs auxiliary equipments in assembly tasks and, therefore, is not suited for applications in unstructured environments.

Examples of research work in the early days include that by Fujii and

Kurono [1], Nakano *et al.* [2], and Takase *et al.* [3]. Already in those pieces of work have been discussed important key issues in the control of multi-arm robots: master/slave control, force/compliance control, and task space control. Nakano *et al.* [2] proposed master/slave force control for the co-ordination of the two arms to carry an object cooperatively. They pointed out the necessity of force control for cooperative multiple robots. The force control is discussed also in [4]. Kurono presented the master/slave control in [5] earlier than Nakano *et al.* [2], incidentally. Fujii and Kurono's proposal in [1] is compliance control for the coordination; they defined a task vector with respect to the object frame and controlled the compliance that was expressed in the coordinate frame. Interesting features in the work by Fujii and Kurono [1] and also by Takase *et al.* [3], by the way, are that both of the work implemented force/compliance control without using any force/torque sensors; they exploited the back-drivability of the actuators. The importance of this technique in practical applications, however, was not recognized at that time. More complicated techniques to use precise force/torque sensors lured people in robotics.

In the 1980's, having had theoretical results for the single-arm robot, strong research on the multi-arm robot was renewed [6]. Definition of task vectors with respect to the object to be handled [7], dynamics and control of the closed-loop system formed by the multi-arm robot and the object [8], [9], and force control issues such as hybrid position/force control [10], [11] have been explored. Through the research work, strong theoretical background for the control of the multi-arm robot is being formed, as is described below, and giving basis for research on more advanced topics, such as cooperative control of dual flexible arms, or development of practical implementation.

How to parameterize the constraint forces/moments on the object, based on the dynamic model for the closed-loop system, is an important issue to be studied; the parameterization gives a task vector for the control and, hence, an answer to one of the most frequently asked questions in the field of multi-arm robotics, that is, how to control simultaneously the trajectory of the object, the contact forces/moments on the object, the load sharing among the arms, and even the external forces/moments on the object.

Many researchers have challenged solving the problem; force decomposition may be a key to solving the problem and has been studied by Uchiyama and Dauchez [12], [13], Walker *et al.* [14], and Bonitz and Hsia [15]. Parameterization of the internal forces/moments on the object to be intuitively understood is important. Williams and Khatib have given a solution to this [16]. Cooperative control schemes based on the parameterization are then designed; they include hybrid control of position/motion and force [12], [13], [17], [18], [19], and impedance control [20], [21].

Load sharing among the arms is also an interesting issue on which many papers have been published [22], [23], [24], [25], [26]. The load sharing is for optimal distribution of the load among the arms. Also, it may be exploited for robust holding of the object when the object is held by the arms without being grasped rigidly. In both cases, anyhow, it becomes a problem of optimization and can be solved by either heuristic methods [27] or mathematical methods [28], [29].

In practical implementation, sophisticated equipments such as force/torque sensors tend to be avoided in industry by many reasons: unreliability, expensiveness, etc. Rebirth of the early methods by Fujii and Kurono [1], or by Inoue [30], should be attractive for people in industry. Hybrid position/force control without using any force/torque sensors but using motor currents at the joints is being successfully implemented in [31]. A key technique in the work is compensation of the friction at the joints.

Recent research is focused on more advanced topics such as handling of multi-bodied objects, or even flexible objects [32], [33], [34], [35]. Also cooperative control of multi-flexible-arm robots is an advanced topic of interest [36], [37]. Once modeling and control problem is solved, the flexible-arm robot is a robot with many merits [38]: it is of light-weight, compliant, and hence safe, etc. Robust holding of the object in presence of slippage of end-effectors on the object may be achieved if the slippage is detected correctly [39].

The rest of the chapter is organized as follows: In Section 1.2, dynamics formulation of closed-loop systems consisting of a multi-arm robot and an object is presented. In Section 1.3, the constraint forces/moments on the object derived in Section 1.2, are elaborated; they are parameterized by external and internal forces/moments. In Section 1.4, a hybrid position/force control scheme that is based on the results in the previous section, is presented, followed by load sharing control methods discussed in Section 1.5. Consideration on practical implementation is given in Section 1.6. Advanced topics being presented in Section 1.7 are mainly those of research in the author's laboratory. This chapter is finally concluded in Section 1.8.

1.2 Dynamics of multi-arm robots

Let suppose the situation depicted in Figure 1.1 where two arms hold a single object. The arms and the object form a closed kinematic chain and, therefore, equations of motion for the system is easily obtained. A point here is that the system is an over-actuated system where the number of actuators to drive the system is more than the number of degrees of freedom of the system. Therefore, how to deal with the constraint forces/moments acting on the system becomes crucial. Here, we formulate those as the

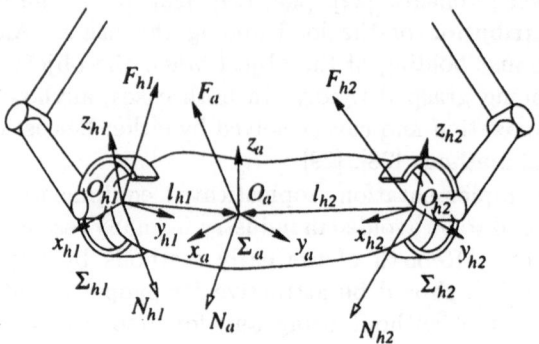

Figure 1.1: Two arms holding an object.

forces/moments that the arms impart to the object.

A model for the analysis that we introduce here is a lumped-mass model and a concept of virtual stick. The virtual stick concept was originally presented in kinematics formulation [12], [13]. The object is modeled as a point with mass and moment of inertia, and the two arms holds the point through the virtual sticks. The point has the same mass and moment of inertia as the object and is located on the center of mass. The model is illustrated in Figure 1.2 with definitions of the frames Σ_a and Σ_i that will be used later in this chapter. With this modeling the formulation becomes straightforward.

Let denote the forces and moments at the point acting on the object through the arm i as $\boldsymbol{f}_i$, then, the forces and moments reacting on the arm through the object is $-\boldsymbol{f}_i$, and the equations of motion of the arm i is given by

$$\boldsymbol{M}_i(\boldsymbol{\theta}_i)\,\ddot{\boldsymbol{\theta}}_i + \boldsymbol{G}_i(\boldsymbol{\theta}_i, \dot{\boldsymbol{\theta}}_i) = \boldsymbol{\tau}_i + \boldsymbol{J}_i^T(\boldsymbol{\theta}_i)\,(-\boldsymbol{f}_i) \tag{1.1}$$

where $\boldsymbol{\theta}_i$ is a vector of the joint variables, $\boldsymbol{\tau}_i$ is a vector of the joint torques or forces, $\boldsymbol{M}_i(\boldsymbol{\theta}_i)$ is an inertia matrix, $\boldsymbol{G}_i(\boldsymbol{\theta}_i, \dot{\boldsymbol{\theta}}_i)$ represents the joint torques or forces due to the centrifugal, Coriolis, gravity, and friction torques or forces at the joints. $\boldsymbol{J}_i(\boldsymbol{\theta}_i)$ is the Jacobian matrix to transform the joint velocity $\dot{\boldsymbol{\theta}}_i$ into the velocity of the frame Σ_i at the tip of the virtual stick.

Another factor to influence the dynamics of the system is that of the object which in this case is obtained as one for a rigid body. Supposing the position and orientation of the object be represented by a vector $\boldsymbol{p}_a$, we have the following equation of motion:

$$\boldsymbol{M}_o(\boldsymbol{\phi})\,\ddot{\boldsymbol{p}}_a + \boldsymbol{G}_o(\boldsymbol{\phi}, \dot{\boldsymbol{\phi}}) = \boldsymbol{f}_1 + \boldsymbol{f}_2 \tag{1.2}$$

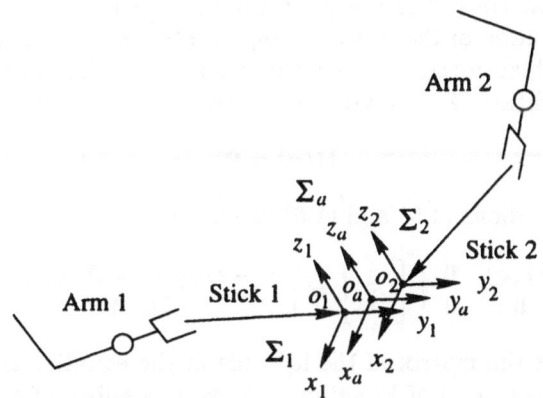

Figure 1.2: A lumped-mass model with virtual sticks.

where ϕ is a vector to represent orientation angles of the object, $M_o(\phi)$ is an inertia matrix of the object, and $G_o(\phi, \dot{\phi})$ represents nonlinear components of the inertial forces such as gravity, centrifugal, and Coriolis forces.

The geometrical constraints imposed on the system come from the fact that the two arms hold the object. Let denote the position and orientation of the object calculated from the joint vector of the arm i as p_i, and suppose that the vector is given by

$$p_i = H_i(\theta_i). \qquad (1.3)$$

Since the object is rigid, the constraints are represented by

$$p_a = H_1(\theta_1) = H_2(\theta_2) \qquad (1.4)$$

where p_a represents the position and orientation of the object.

Now, we have a set of fundamental equations to describe the dynamics of the closed-loop system, that consists of the differential equations (1.1) and (1.2) to describe the dynamics of the arms and the object, respectively, and the algebraic equation (1.4) to represent the constraint condition.

The system of equations forms a singular system and the solution is obtained as follows [8]: The differential equations (1.1) and (1.2) are written by one equation as

$$M(q)\ddot{q} + G(q, \dot{q}) = \tau + J^T(q)\lambda \qquad (1.5)$$

where $M(q)$ is the inertia matrix of the whole system, $G(q, \dot{q})$ represents the nonlinear components of the whole system, q is a vector of generalized coordinates that consist of the joint variables of the arms and the position and orientation of the object, τ represents the generalized forces, and $J(q)$ is a Jacobian matrix. λ represents constraint forces/moments. The constraint condition (1.4) is written in a compact form as

$$H(q) = 0. \tag{1.6}$$

Combining Equations (1.5) and (1.6), we have

$$\begin{bmatrix} M(q) & 0 \\ 0 & 0 \end{bmatrix} \begin{bmatrix} \ddot{q} \\ \dot{\lambda} \end{bmatrix} = \begin{bmatrix} \tau - G(q, \dot{q}) + J^T(q)\lambda \\ H(q) \end{bmatrix}. \tag{1.7}$$

It is noted that the matrix in the left side of the equation is singular and hence direct integration of Equation (1.7) is impossible, of course.

The solution of Equation (1.7) is obtained after the reduction transformation as follows [8]: Differentiating the constraint condition twice by time, we have

$$\ddot{H}(q) = J(q)\ddot{q} + \dot{J}(q)\dot{q} = 0. \tag{1.8}$$

Since $M(q)$ in Equation (1.5) is positive definite, its inverse exists and we have

$$\ddot{q} = M(q)^{-1} \left\{ \tau + J^T(q)\lambda - G(q, \dot{q}) \right\}. \tag{1.9}$$

Substituting Equation (1.9) into Equation (1.8), we have

$$J(q)M(q)^{-1}J^T(q)\lambda = J(q) \left[M(q)^{-1} \left\{ G(q, \dot{q}) - \tau \right\} \right] - \dot{J}(q)\dot{q}. \tag{1.10}$$

Therefore,

$$\lambda = \left\{ J(q)M(q)^{-1}J^T(q) \right\}^{-1} \left\{ J(q) \left[M(q)^{-1} \left\{ G(q, \dot{q}) - \tau \right\} \right] - \dot{J}(q)\dot{q} \right\}. \tag{1.11}$$

From Equations (1.9) and (1.11), we obtain q and λ, that is the solution for a given τ.

It is noted that the inverse kinematics problem of flexible-arm robots is formulated as a problem of finding a solution for a set of differential-algebraic equations [40]. The problem may be mathematically similar to the one in this section.

1.3 Derivation of task vectors

The task vector consists of a set of variables that is convenient for describing a given task. A set of Cartesian coordinates in the workspace forms a task vector for a task of carrying an object in the workspace, for example. For more complicated tasks that include constrained motion, it has to be defined not only as position/orientation of the object but also as forces/moments acting on the object. In this section, we derive task vectors to describe a task performed by the multi-arm robot.

The constraint forces/moments f_i are those applied to the object by the arm i and are obtained from Equation (1.11) when the joint torques or forces τ_i are given. Since f_i is 6-dimensional, the forces/moments applied to the object by the two arms are altogether 12-dimensional, six of which are for driving the object, and the rest of which do not contribute to the motion of the object but yield internal forces/moments on the object. Noting this intuition, we derive the task vector for the cooperative two arms [12], [13], [24].

1.3.1 External and internal forces/moments

First, the external forces/moments on the object are defined as those to drive the object. That is,

$$
\begin{aligned}
f_a &= f_1 + f_2 \\
&= [I_6 \ \ I_6] \begin{bmatrix} f_1^T & f_2^T \end{bmatrix}^T \\
&= W\lambda
\end{aligned}
\tag{1.12}
$$

where W is a 6×12 matrix with range of 6-dimension and null space of 6-dimension. I_n is the unit matrix of n-dimension. This relation is shown in Figure 1.3 (a). A solution for λ when f_a is given is

$$
\begin{aligned}
\lambda &= W^+ f_a + (I_{12} - W^+ W)\, z \\
&= W^+ f_a + [I_6 \ \ -I_6]^T f_r \\
&= W^+ f_a + V f_r
\end{aligned}
\tag{1.13}
$$

where W^+ is the Moore-Penrose inverse of W given by

$$
W^+ = \begin{bmatrix} \frac{1}{2} I_6 \\ \frac{1}{2} I_6 \end{bmatrix}.
\tag{1.14}
$$

z is an arbitrary vector of 12-dimension. The second term of the right hand side of Equation (1.13) represents the null space of W, and V represents its bases by which the vector f_r is represented. The relation is shown in Figure 1.3 (b). It is apparent when viewing V that f_r represents forces/moments being applied by the two arms in opposite directions.

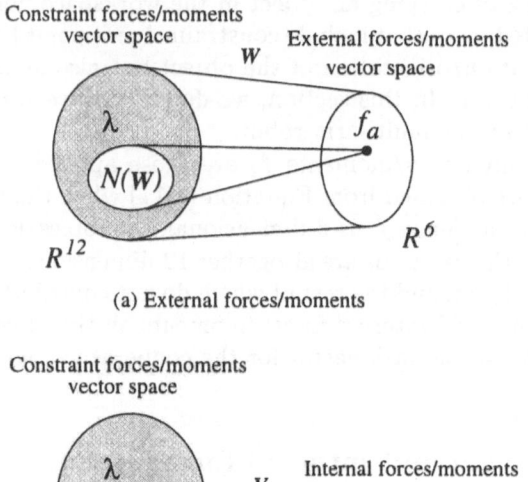

(a) External forces/moments

(b) Internal forces/moments

Figure 1.3: External and internal forces/moments.

We call the forces/moments represented by f_r internal forces/moments. Solving Equation (1.13) for f_a and f_r, we have

$$f_a = f_1 + f_2 \qquad \qquad \cdot \; (1.15)$$

$$f_r = \frac{1}{2}(f_1 - f_2). \qquad \qquad (1.16)$$

1.3.2 External and internal velocities

The velocities corresponding to the external and internal forces/moments are derived using the principle of virtual work, as follows:

$$s_a = \frac{1}{2}(s_1 + s_2) \qquad \qquad (1.17)$$

$$\Delta s_r = s_1 - s_2 \qquad \qquad (1.18)$$

where s_a, Δs_r, s_1 and s_2 are velocity vectors corresponding to f_a, f_r, f_1 and f_2, respectively. The velocities s_a, s_1 and s_2 are those of Σ_a, Σ_1 and Σ_2 in Figure 1.2, respectively.

1.3.3   External and internal positions/orientations

The positions/orientations corresponding to external and internal forces/moments are derived by integrating the relation in Equations (1.17) and (1.18), as follows:

$$p_a = \frac{1}{2}(p_1 + p_2) \tag{1.19}$$

$$\Delta p_r = p_1 - p_2 \tag{1.20}$$

where p_a, Δp_r, p_1 and p_2 are position/orientation vectors corresponding to s_a, s_r, s_1 and s_2, respectively. The positions/orientations p_a, p_1 and p_2 are those of Σ_a, Σ_1 and Σ_2 in Figure 1.2, respectively.

An alternative way of representing the positions/orientations is to use the homogeneous transformation matrix [24]: The positions and orientations of the frames Σ_1 Σ_2 in Figure 1.2 is represented by

$$H_i = \begin{bmatrix} n_i & o_i & a_i & x_i \\ 0 & 0 & 0 & 1 \end{bmatrix}. \tag{1.21}$$

Corresponding to the positions/orientations p_a and Δp_r, the homogeneous transformation matrix to represent the position/orientation of the frame Σ_a:

$$H_a = \begin{bmatrix} n_a & o_a & a_a & x_a \\ 0 & 0 & 0 & 1 \end{bmatrix} \tag{1.22}$$

and the vectors Δx_r, $\Delta \Omega_r$ to represent the small (virtual) deformation of the object are derived as follows:

$$n_a = \frac{1}{2}(n_1 + n_2) \tag{1.23}$$

$$o_a = \frac{1}{2}(o_1 + o_2) \tag{1.24}$$

$$a_a = \frac{1}{2}(a_1 + a_2) \tag{1.25}$$

$$x_a = \frac{1}{2}(x_1 + x_2) \tag{1.26}$$

$$\Delta x_r = x_1 - x_2 \tag{1.27}$$

$$\Delta \Omega_r = \frac{1}{2}(n_2 \times n_1 + o_2 \times o_1 + a_2 \times a_1). \tag{1.28}$$

1.4 Hybrid position/force control

In the previous section we have seen that the task vectors for the cooperative two arms are the external and internal forces/moments, velocities,
and positions/orientations. The internal positions/orientations are constrained in tasks such as carrying a rigidly held object. Therefore, a certain force-related control scheme should be applied to the control of the
cooperative two arms. There have been proposed different schemes for the
force-related control. They include compliance control [1], hybrid control of
position/motion and force [10], [11], [12], [13], [17], [18], [19], and impedance
control [20], [21]. Any of those control schemes will be successfully applied
to the control of this system if the task vector is properly chosen. For those
systems that this chapter deals with and in which constraint conditions are
clearly stated, however, hybrid position/force control will be most suitably
used. The rest of this section, therefore, describes the hybrid position/force
control [12], [13], [17].

Using the equations derived in Section 1.3, the task vectors for the
hybrid position/force control are defined as

$$z = \begin{bmatrix} p_a^T & \Delta p_r^T \end{bmatrix}^T \tag{1.29}$$

$$u = \begin{bmatrix} s_a^T & \Delta s_r^T \end{bmatrix}^T \tag{1.30}$$

$$h = \begin{bmatrix} f_a^T & f_r^T \end{bmatrix}^T \tag{1.31}$$

where z, u, and h are the task position, velocity, and force vectors, respectively. The organization of the control scheme is shown in Figure 1.4,
diagrammatically. The suffixes r, c and m represent the reference value,
current value and control command, respectively. The command vector e_r
to the actuators of the two arms is calculated by

$$e_r = e_z + e_h \tag{1.32}$$

where e_z is the command vector for the position control and is calculated
by

$$e_z = K_z J_\theta^{-1} G_z(s) S B_a (z_r - z_c) \tag{1.33}$$

and e_h is the command vector for the force control and is calculated by

$$e_h = K_h J_\theta^T G_h(s)(I - S)(h_r - h_c). \tag{1.34}$$

B_a in Equation (1.33) is a matrix to transform the errors of orientation
angles into a rotation vector. J_θ is the Jacobian matrix to transform the
joint velocity $\dot{\theta} = [\dot{\theta}_1^T \ \dot{\theta}_2^T]^T$ into the task vector of velocity u. $G_z(s)$

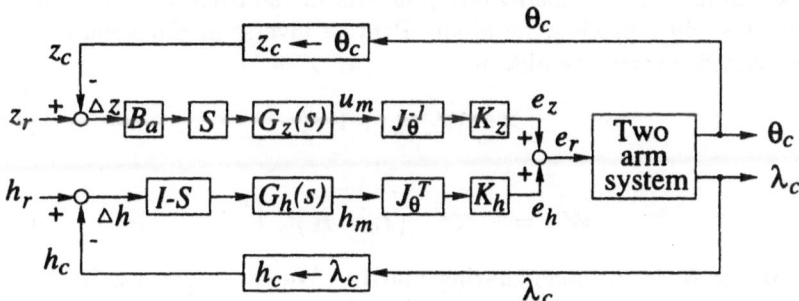

Figure 1.4: A hybrid position/force control scheme.

and $G_h(s)$ are operator matrices representing position and force control laws, respectively. The matrices K_z and K_h are assumed to be diagonal. Their diagonal elements convert velocity and force commands into actuator commands, respectively. S is a matrix to switch the control modes from position to force or vice versa. S is diagonal and its diagonal elements take the values of 1 or 0. The ith workspace coordinate is position-controlled if the ith diagonal element of S is 1, and force-controlled if 0. I is the unit matrix with the same dimension as S. θ_c and λ_c are vectors of measured joint variables and measured forces/moments, respectively.

In the above control scheme, without distinguishing a master nor a slave, the two arms are controlled cooperatively. It is not necessary to assign master and slave modes to each arm. Also, in the control of internal forces/moments, since the references to the external positions/orientations are sent to the both arms, the disturbance from the position control loop to the force control loop is decreased. This enables the above scheme to achieve more precise force control than the master/slave scheme [11], [17].

1.5   Load sharing

The problem of load sharing in the multi-arm robot system is that of how to distribute the load to each arm; a strong arm may share the load more than a weak one, for instance. This is possible because the multi-arm robot has redundant actuators; if the robot has only sufficient number of actuators for supporting the load, no optimization of load distribution is possible. In this section, we elaborate this problem according to our previous work [24], [27], [28], [29]. Also, it should be noted that the work by Unseren [25], [26] is more comprehensive.

We can introduce a load-sharing matrix in the framework presented in Section 1.3. By replacing the Moore-Penrose inverse in Equation (1.13) by a generalized inverse, we obtain:

$$\lambda = W^- f_a + V f'_r \tag{1.35}$$

where

$$W^- = \left[\begin{array}{cc} K^T & (I_6 - K)^T \end{array} \right]^T. \tag{1.36}$$

The matrix K is the load-sharing matrix. We can prove easily that the non-diagonal elements of K only yield a λ vector in the null space of W, that is, the space of internal forces/moments. Therefore, without loosing generality, let us choose K such that:

$$K = \text{diag} \left[\alpha_i \right] \tag{1.37}$$

where we call α_i a load sharing coefficient.

Now, the problem we have to deal with is that of how to tune the load sharing coefficient α_i to ensure correct manipulation of the object by the two arms. To answer this question, we have to notice first that by mixing Equations (1.13) and (1.35), we obtain:

$$f_r = V^{-1} \left(W^- - W^+ \right) f_a + f'_r \tag{1.38}$$

which, by recalling that only f_a and λ are really existing forces/moments, notifies that:

- f_r, f'_r and α_i are "artificial" parameters introduced for better understanding of the manipulation process.

- f'_r and α_i are not independent quantities. The concept of internal forces/moments and the concept of load sharing are mathematically mixed with each other.

Therefore, we can conclude that to tune the load sharing coefficients or to choose suitable internal forces/moments is strictly equivalent from the mathematical and also from the performance point of view. One of f_r, f'_r and α_i is independent parameters, that is redundant parameters, to be optimized for load sharing. This is more generally stated in [25], [26]. We have proposed to tune the internal forces/moments f_r for simplicity of equations and also for consistency with control [28], [29].

One of interesting problems regarding the load sharing is that of robust holding: a problem to determine the forces/moments λ, which the two arms apply to the object, in order not to drop it even when disturbing external forces/moments are applied to it. This problem can be solved by tuning the internal forces/moments (or the load sharing coefficients, of course).

This problem is addressed in [27], where conditions to keep holding are expressed by the forces/moments at the end-effectors, and Equation (1.35) being substituted into the conditions, a set of linear inequalities for both f_r' and α_i are obtained as:

$$A f_r' + B\alpha < c \qquad (1.39)$$

where A and B are 6×6 matrices, c a 6-dimensional vector, and $\alpha = [\alpha_1, \ \alpha_2, \ \cdots, \ \alpha_6]^T$. In the paper [27], a solution of α_i for the inequality is obtained, heuristically. The above inequality can be transformed into that with respect to f_r, of course, but the parameter α_i is fitter to such heuristic algorithm because α_i can be understood intuitively.

The same problem may be solved mathematically: introducing an objective function to be optimized, we can formulate the problem as that of mathematical programming. For that purpose, we choose a quadratic function of f_r as

$$\min f_r^T Q f_r \qquad (1.40)$$

where Q is a 6×6 positive definite matrix. The objective function represents a kind of energy to be consumed by the joint actuators; the arms consume electric energy in the actuators to yield the internal forces/moments f_r. The problem to minimize the objective function under the constraints is a quadratic programming problem. A solution can be found in [28], [29].

1.6 Practical implementation

The results regarding the topics of this chapter that the researchers have yielded so far are of value, of course, but not being used in industry. Why are they not being used? A reason will be that the schemes require sophisticated force/torque sensors and special control software that is incompatible to current industrial robots. Therefore, we re-examine the control scheme that we presented in the previous sections so that it may be used in industry. A solution will be that we should use motor currents to get force/torque information instead of wrist force/torque sensors. This technique was proposed first long time ago [30].

To see if this solution is feasible, we implement the hybrid position/force control scheme in Section 1.4 by a robot developed for experimental research on application to shipbuilding work [31]. The robot is called C-ARM (Cooperative-ARM) and is drawn in Figure 1.5. Each arm has three degrees-of-freedom in the same vertical plane. The first joint is a prismatic

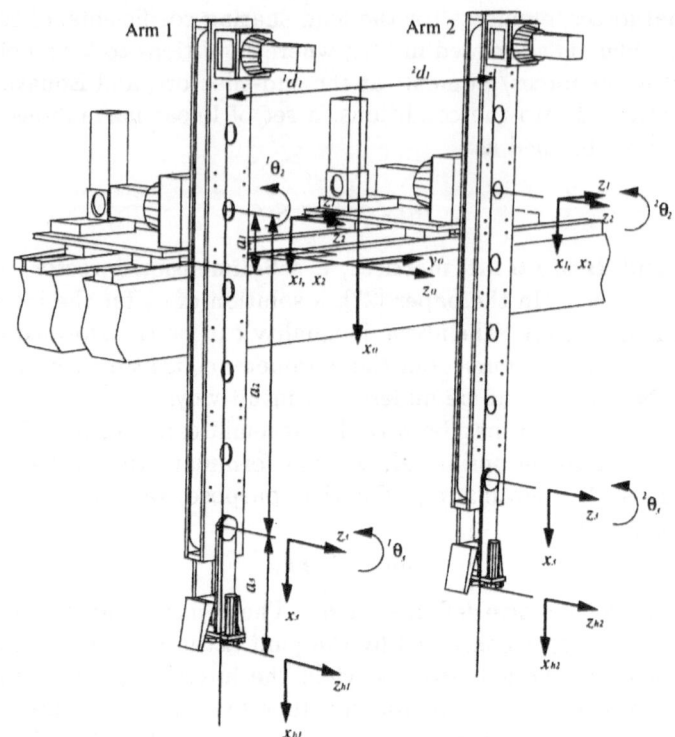

Figure 1.5: An experimental two-arm robot.

one, and the second and the third are rotary. Every joint is actuated by an AC servo motor. The torque τ_{mi} at the ith joint is proportional to the motor current I_{ai}, that is:

$$\tau_{mi} = K_{ti} I_{ai} \tag{1.41}$$

where K_{ti} is a constant of proportionality for the motor. Using this property, we realize cooperative control without using any force/torque sensors but measuring the motor currents only.

Generally, the motor torques at each joint are amplified by the reduction gears. The amplified torques are represented by

$$K_r \tau_m = \tau_p + \tau_f \tag{1.42}$$

where K_r is a diagonal matrix with the reduction gear ratios as its diagonal elements, τ_m is a motor torque vector, τ_p is a torque vector to move the arms, and τ_f is a torque vector corresponding to the forces/moments on

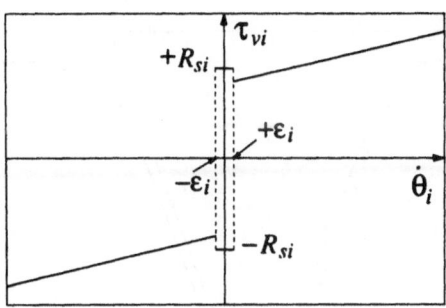

Figure 1.6: A friction model.

the end-effectors. The term $K_r \tau_m$ is obtained by measuring the motor currents.

Now, a question arises: how can we get only τ_f from the measured values of the motor currents? To answer this, we need to derive equations of motion for the robot with considering even the dynamics of its motors and the reduction gears. Neglecting the inertia terms due to the link dynamics in the equations of motion, we obtain

$$\tau_p = \tau_v + \tau_g \tag{1.43}$$

where τ_v and τ_g are the friction and the gravity force vectors, respectively. Then, τ_p is deducted from the joint torque $K_r \tau_m$, to yield τ_f, which in turn is used to calculate the forces/moments being applied to the object, and hence external and internal forces/moments on the object.

The friction model used for the calculation of τ_v is shown in Figure 1.6. This model includes both Coulomb and viscous frictions. In Figure 1.6, θ_i and R_{si} are the joint angle and the maximum static friction, respectively, while ε_i is a constant parameter which is a threshold for the approximation of $\dot{\theta}_i$ to be zero. Here, since it is difficult to decide the direction and magnitude of the static friction, we propose a novel method in which static friction is switched alternately between $+R_{si}$ and $-R_{si}$ at each sampling period Δt. In this case, the friction of each joint τ_{vi} is written as follows:

$$\tau_{vi} = \begin{cases} +R_{si} \overset{\Delta t}{\leftrightarrow} -R_{si} & -\varepsilon_i \leq \dot{\theta}_i \leq \varepsilon_i \\ V_i \dot{\theta}_i + R_{ci} \, \text{sgn}(\dot{\theta}_i) & \dot{\theta}_i < -\varepsilon_i \quad \varepsilon_i < \dot{\theta}_i \end{cases} \tag{1.44}$$

where V_i and R_{ci} are the coefficients of viscous and Coulomb frictions, respectively.

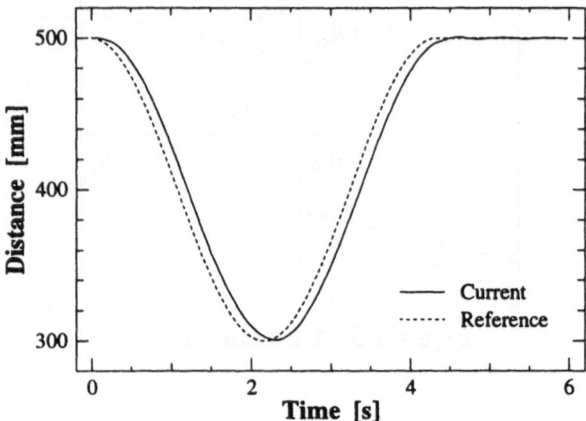

Figure 1.7: Experimental results: ^0x_a.

On applying the hybrid position/force control scheme, we find the robot has three external and three internal degrees of freedom. In the experiment, the external and internal coordinates are controlled by position- and force-control modes, respectively. For reference positions/orientations, those at the object center is used, and for reference forces/moments, the internal forces/moments on the object are taken.

Experimental results are shown in Figures 1.7–1.10. The reference positions/orientations are given as a cosine curve in x direction, and the reference internal forces/moments are all set to be constant. To see if the internal forces/moments calculated from the motor currents give good estimation, we compared those with the internal forces/moments measured by a force/torque sensor embedded in the object. The "Sample" in Figure 1.9 and 1.10 means that the force data are obtained by the force/torque sensor. As to the internal forces, large vibrations in the controlled values are observed when the object velocity approaches to zero. These vibrations are caused by the static friction determined by Equation (1.44). The forces measured by the force/torque sensor, however, are held close to the reference values during the whole task. This shows that our method works well.

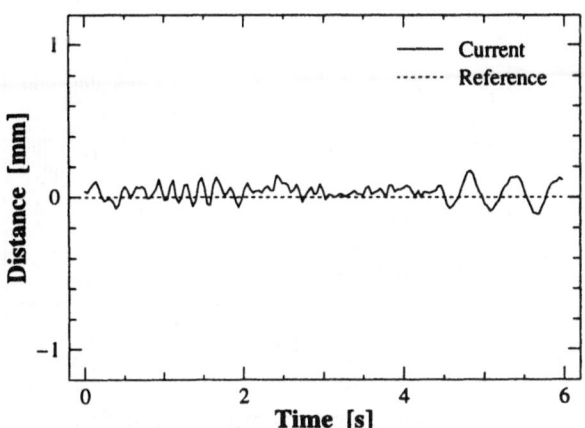

Figure 1.8: Experimental results: ^0y_a.

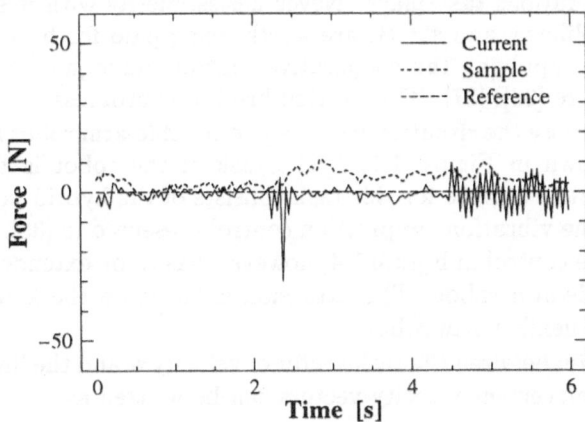

Figure 1.9: Experimental results: $^aF_{rx}$.

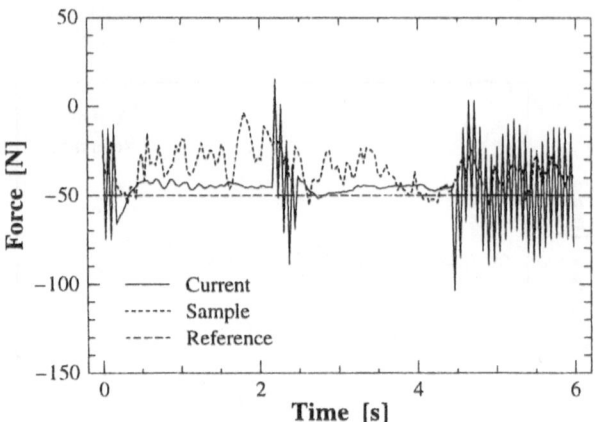

Figure 1.10: Experimental results: $^aF_{ry}$.

1.7 Advanced topics

1.7.1 Multi-flexible-arm robots

The flexible-arm robot is a robot with light weight and structural compliance. Due to the compliance, demerits such as positioning errors and structural vibrations take place. Nevertheless, merits with it such as light weight, compliance, and safety, are worth being paid for by the disadvantages. We are applying the cooperative control techniques to this kind of robots of future [36], [37]. This section briefs our progress.

Let us suppose the situation where a two-flexible-arm robot hold a single object as shown in Figure 1.11. The task of the robot is to carry the object. A control scheme for this task consists of the hybrid position/force control and the vibration-suppression control presented in [38]. The hybrid position/force control in Figure 1.4, however, has to be extended to the one for the flexible-arm robot. The extension is based on the kinematics and statics of the flexible-arm robot.

The relation between the task vector of velocity u and the link-deflection and joint-displacement velocity vectors can be written as:

$$u = J_\theta \dot{\theta} + J_e \dot{e} \tag{1.45}$$

where $\theta = [\theta_1^T \ \theta_2^T]^T$ and $e = [e_1^T \ e_2^T]^T$ are the joint-displacement and the link-deflection vectors, respectively. Since $\dot{e}$ are kept small during the

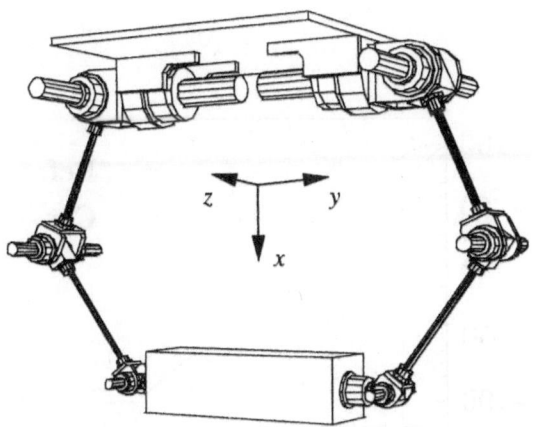

Figure 1.11: Overview of ADAM handling a rigid object.

vibration control, $J_e\dot{e}$ can be neglected. Therefore, the joint-displacement-velocity vector $\dot{\theta}_z$ for position control is calculated from the task vector of velocity u_m as:

$$\dot{\theta}_z = J_\theta^{-1} u_m. \qquad (1.46)$$

Considering Equations (1.30) and (1.31) and the duality relation between forces and velocities, the vector of joint torques $\tau_h = [\tau_{h1}^T \ \tau_{h2}^T]^T$ for force control is calculated from the force/moment vector h_m as follows:

$$\tau_h = J_\theta^T h_m. \qquad (1.47)$$

By using Equations (1.46) and (1.47), the hybrid position/force control scheme for the flexible-arm robot can be organized like for the rigid one.

In addition to the above control, vibration-suppression control has to be included for stable cooperation. In the study, the scheme in [38] is applied to each arm. The mass of the object is neglected.

To illustrate the validity of the proposed scheme, experiments on a real flexible robot with two flexible arms having seven degrees-of-freedom for each are performed. The robot is called ADAM (Aerospace Dual-Arm Manipulator) [41] and has been shown in Figure 1.11. In the experiment, a light and rigid object is handled. Results of the experiments are shown in Figures 1.12–1.15. In the particular experiment, the reference positions/orientations and forces/moments but the position of z are kept constant. Figure 1.12 shows the position of the object, while Figures 1.13, 1.14, and 1.15 show the internal forces exerted on the object. The experimental results illustrate the effectiveness of the proposed control scheme.

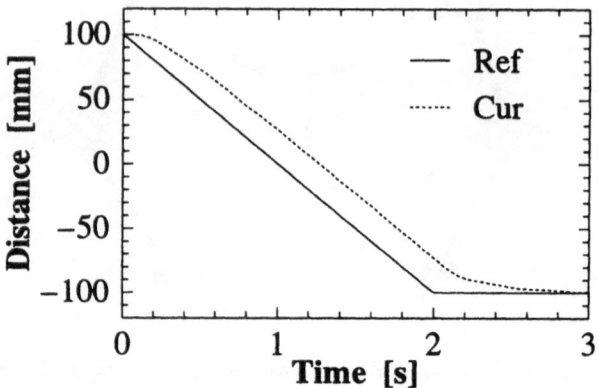

Figure 1.12: Experimental results: ^0z_a.

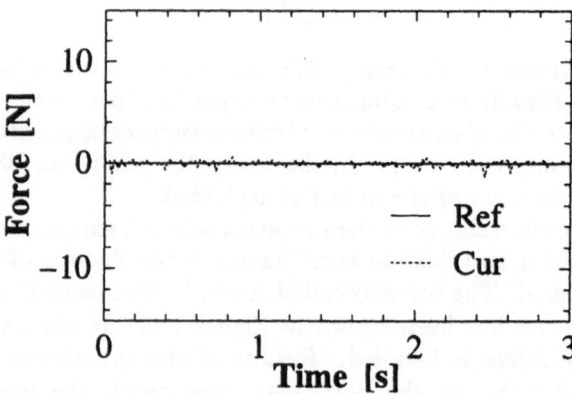

Figure 1.13: Experimental results: $^aF_{rx}$.

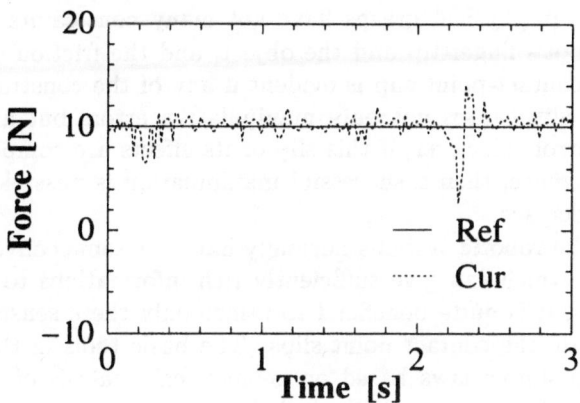

Figure 1.14: Experimental results: $^{a}F_{ry}$.

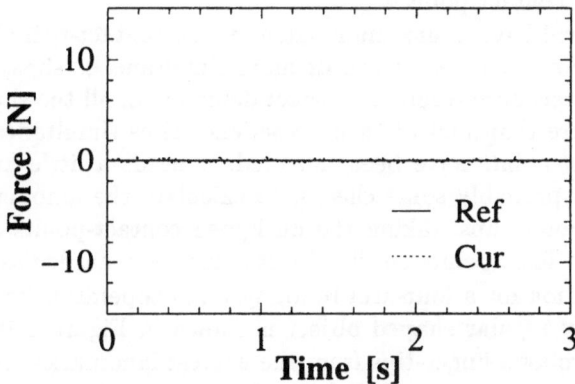

Figure 1.15: Experimental results: $^{a}F_{rz}$.

1.7.2   Slip detection and robust holding

Cooperating multiple robots experience slip when grasps on the object are defined by the internal forces developed due to each robot. Such manipulations without physical grasps have got many constraints like friction between a robot's finger-tip and the object, and the friction cone defined due to it. A contact-point slip is evident if any of the constraints is overlooked. This slip causes not only manipulation errors but also a failure of system control. However, if this slip or its effects are compensated just after its occurrence, then a successful manipulation is possible even in an enhanced workspace.

Since all the robotic systems normally have got some conventional and cheap sensors which can give sufficiently rich informations to localize the end-point tips, it is quite beneficial to utilize only these sensors to detect and compensate the contact-point slips. The basic tools in this approach are some very simple laws based on geometrical analysis of the mesh of links developed by inter-connecting all the contact-points. The main tool is a slip indicator S_i, which is defined as

$$S_i = \sum_{j=1}^{n} \mid \Delta R_{ij} \mid \qquad (1.48)$$

where $i = j = 1, 2, 3, \cdots, n$ is the contact-point number. ΔR_{ij} is the change in an inter-contact link between ith and jth contact-points after a slip occurs. S_i sums up all these absolute changes for the links having their one end at ith contact-point.

Surely S_i will have a maximum value for the contact-point which actually slips. For a few cases of two or more simultaneous slips, a recursion in the above procedure results in correct detection of all the slipped finger-tips unless more than half of them experience slips simultaneously. Once slipped contact-points have been detected, it needs a little knowledge of geometry, and probably some checks, to calculate the amounts by which each contact-point slips, taking the unslipped contact-points as reference and some other fixed points on the object's surface, regarded as landmarks.

An illustration for a four-arm robot system cooperating to manipulate a geometrically regular shaped object is shown in Figure 1.16. The distances of the robot's finger-tips from the nearest landmarks are defined as α_i. These distances are very helpful in determining the physical amounts of slips geometrically. The control structure which takes into account the phenomenon of slip is shown in Figure 1.17. This control method generates the actuator force commands for a proper force distribution between all the arms to generate a resultant external force corresponding to the

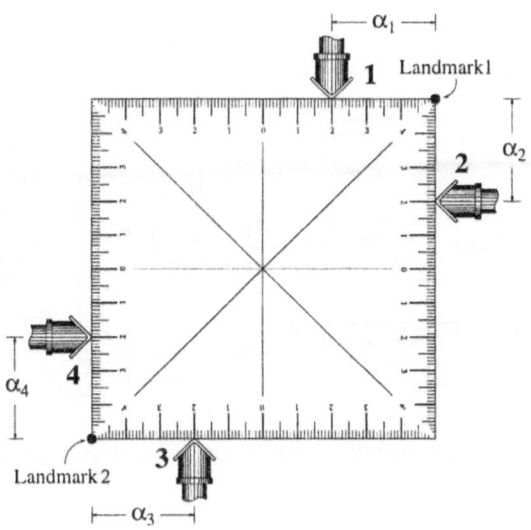

Figure 1.16: Four-arm robot system cooperating at an object.

desired manipulation along with maintaining certain fixed internal forces responsible for grasps.

The experimental results obtained using the control algorithm of Figure 1.17 on the system of Figure 1.16 are shown in Figures 1.18–1.21. For a manipulation with no slip, all the values of α_i should remain constant. But as a finger slips, the new values of α_i are calculated after an execution of the slip detection algorithm for all manipulating arms. The results show that a successful object manipulation was possible even after two contact-points changed their positions due to occurrence of slips at different time intervals.

A sensor-based approach is to employ a vision-tracking system for slip detection. One way is to track the contact-points and whenever there occurs a slip, its amount is known by making a comparison with previously tracked video frames, while the other way is to track the object being manipulated; in this way the vision-tracking system acts as a sensor for the actual posture of the object. This approach should work well as a sensor for object's posture is present in the main control loop, but the main problem is the slow tracking speed which is dependent on video scanning speed. Moreover, another problem is the high cost of this system which makes the overall system a cost non-effective one.

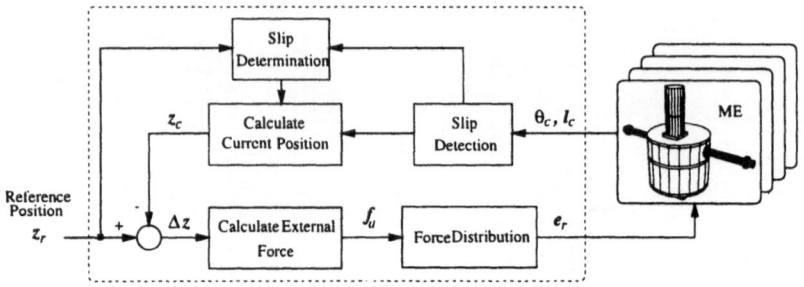

Figure 1.17: Control algorithm considering slip detection/compensation.

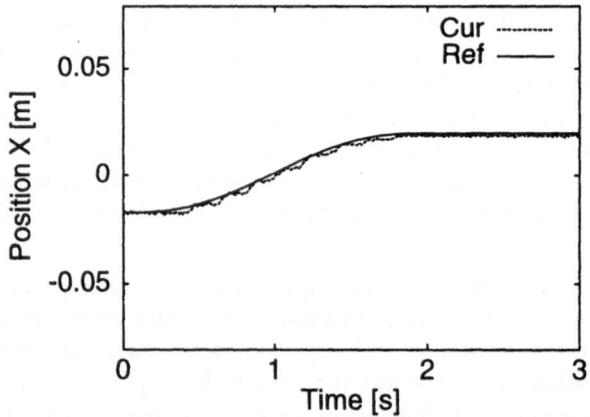

Figure 1.18: Experimental results: Position along x.

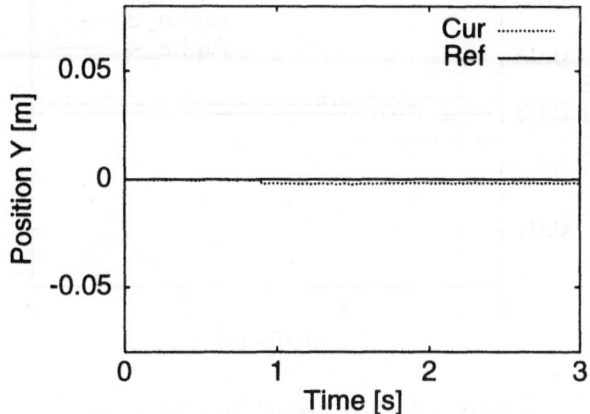

Figure 1.19: Experimental results: Position along y.

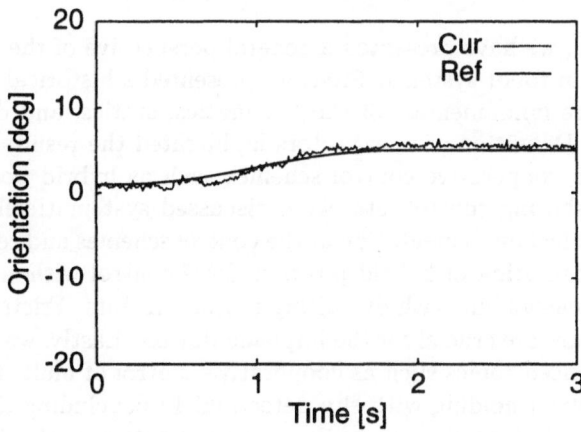

Figure 1.20: Experimental results: Orientation.

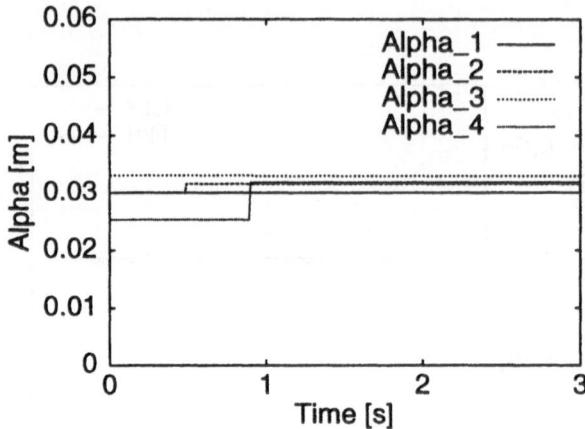

Figure 1.21: Experimental results: α.

1.8   Conclusions

In this chapter, we have presented a general perspective of the state of the art of multi-arm robot systems. First, we presented a historical perspective and, then, gave fundamentals of the kinematics, statics, and dynamics of such systems. Definition of task vectors highlighted the results and gave a basis on which cooperative control schemes such as hybrid position/force control, load sharing control, etc. were discussed systematically. We also discussed practical implementation of the control schemes and reported successful implementation of hybrid position/force control without using any force/torque sensors but with exploiting motor currents. Friction compensation techniques are crucial for the implementation. Lastly, we presented a couple of advanced topics such as cooperative control of multi-flexible-arm robots, and robust holding with slip detection. In concluding this chapter, we should note that application of theoretical results to real robot systems is of prime importance, and that efforts in future research will be directed in this direction to yield stronger results. Advanced topics for future research will include kinematics for more sophisticated tasks [42] and decentralized control [43].

Acknowledgements

The author records acknowledgments to Prof. Kazuhiro Kosuge, Dr. Mikhail M. Svinin, Mr. Yuichi Tsumaki, Mr. Khalid Munawar, Mr. Yoshihiro Tanno, and Mr. Mitsuhiro Yamano who helped him in preparing this chapter.

References

[1] S. Fujii and S. Kurono, "Coordinated Computer Control of a Pair of Manipulators," in *Proc. 4th IFToMM World Congress,* Newcastle upon Tyne, England, September 1975, pp. 411–417.

[2] E. Nakano, S. Ozaki, T. Ishida, and I. Kato, "Cooperational Control of the Anthropomorphous Manipulator 'MELARM'," in *Proc. 4th Int. Symp. Industrial Robots,* Tokyo, Japan, November 1974, pp. 251–260.

[3] K. Takase, H. Inoue, K. Sato, and S. Hagiwara, "The Design of an Articulated Manipulator with Torque Control Ability," in *Proc. 4th Int. Symp. Industrial Robots,* Tokyo, Japan, November 1974, pp. 261–270.

[4] T. Ishida, "Force Control in Coordination of Two Arms," in *Proc. 5th Int. Conf. on Artificial Intelligence,* 1977, pp. 717–722.

[5] S. Kurono, "Cooperative Control of Two Artificial Hands by a Mini-Computer," in *Prepr. 15th Joint Conf. on Automatic Control,* November 1972, pp. 365–366, (in Japanese).

[6] A. J. Koivo and G. A. Bekey, "Report of Workshop on Coordinated Multiple Robot Manipulators: Planning, Control, and Applications," *IEEE J. of Robotics and Automation,* vol. 4, no. 1, pp. 91–93, 1988.

[7] P. Dauchez and R. Zapata, "Co-Ordinated Control of Two Cooperative Manipulators: the Use of a Kinematic Model," in *Proc. 15th Int. Symp. Industrial Robots,* Tokyo, Japan, September 1985, pp. 641–648.

[8] N. H. McClamroch, "Singular Systems of Differential Equations as Dynamic Models for Constrained Robot Systems," in *Proc. 1986 IEEE Int. Conf. on Robotics and Automation,* San Francisco, USA, April 1986, pp. 21–28.

[9] T. J. Tarn, A. K. Bejczy, and X. Yun, "New Nonlinear Control Algorithms for Multiple Robot Arms," *IEEE Trans. on Aerospace and Electronic Systems,* vol. 24, no. 5, pp. 571–583, 1988.

[10] S. Hayati, "Hybrid Position/Force Control of Multi-Arm Cooperating Robots," in *Proc. 1986 IEEE Int. Conf. on Robotics and Automation,* San Francisco, USA, April 1986, pp. 82–89.

[11] M. Uchiyama, N. Iwasawa, and K. Hakomori, "Hybrid Position/Force Control for Coordination of a Two-Arm Robot," in *Proc. 1987 IEEE Int. Conf. on Robotics and Automation,* Raleigh, USA, March 1987, pp. 1242–1247.

[12] M. Uchiyama and P. Dauchez, "A Symmetric Hybrid Position/Force Control Scheme for the Coordination of Two Robots," in *Proc. 1988 IEEE Int. Conf. on Robotics and Automation,* Philadelphia, USA, April 1988, pp. 350–356.

[13] M. Uchiyama and P. Dauchez, "Symmetric Kinematic Formulation and Non-Master/Slave Coordinated Control of Two-Arm Robots," *Advanced Robotics: The International Journal of the Robotics Society of Japan,* vol. 7, no. 4, pp. 361–383, 1993.

[14] I. D. Walker, R. A. Freeman, and S. I. Marcus, "Analysis of Motion and Internal Force Loading of Objects Grasped by Multiple Cooperating Manipulators," *Int. J. of Robotics Research,* vol. 10, no. 4, pp. 396–409, 1991.

[15] R. G. Bonitz and T. C. Hsia, "Force Decomposition in Cooperating Manipulators Using the Theory of Metric Spaces and Generalized Inverses," in *Proc. 1994 IEEE Int. Conf. on Robotics and Automation,* San Diego, USA, May 1994, pp. 1521–1527.

[16] D. Williams and O. Khatib, "The Virtual Linkage: A Model for Internal Forces in Multi-Grasp Manipulation," in *Proc. 1993 IEEE Int. Conf. on Robotics and Automation,* Atlanta, USA, May 1993, pp. 1025–1030.

[17] M. Uchiyama and Y. Nakamura, "Symmetric Hybrid Position/Force Control of Two Cooperating Robot Manipulators," in *Proc. 1988 IEEE Int. Workshop on Intelligent Robots and Systems,* Tokyo, Japan, November 1988, pp. 515–520.

[18] J. T. Wen and K. Kreutz-Delgado, "Motion and Force Control of Multiple Robotic Manipulators," *Automatica,* vol. 28, no. 4, pp. 729–743, 1992.

[19] V. Perdereau and M. Drouin, "Hybrid External Control for Two Robot Coordinated Motion," *Robotica,* vol. 14, pp. 141–153, 1996.

[20] K. Kosuge, M. Koga, and K. Nosaki, "Coordinated Motion Control of Robot Arm Based on Virtual Internal Model," in *Proc. 1989 IEEE Int. Conf. on Robotics and Automation,* Scottsdale, USA, May 1989, pp. 1097–1102.

[21] M. Koga, K. Kosuge, K. Furuta, and K. Nosaki, "Coordinated Motion Control of Robot Arms Based on the Virtual Internal Model," *IEEE Trans. on Robotics and Automation,* vol. 8, no. 1, pp. 77–85, 1992.

[22] D. E. Orin and S. Y. Oh, "Control of Force Distribution in Robotic Mechanisms Containing Closed Kinematic Chains," *Trans. ASME, J. of Dynamic Systems, Measurement, and Control,* vol. 102, pp. 134–141, 1981.

[23] Y. F. Zheng and J. Y. S. Luh, "Optimal Load Distribution for Two Industrial Robots Handling a Single Object," in *Proc. 1988 IEEE Int. Conf. on Robotics and Automation,* Philadelphia, USA, April 1988, pp. 344–349.

[24] M. Uchiyama, "A Unified Approach to Load Sharing, Motion Decomposing, and Force Sensing of Dual Arm Robots," *Robotics Research: The Fifth International Symposium,* Edited by H. Miura and S. Arimoto, The MIT Press, pp. 225–232, 1990.

[25] M. A. Unseren, "A New Technique for Dynamic Load Distribution when Two Manipulators Mutually Lift a Rigid Object. Part 1: The Proposed Technique," in *Proc. First World Automation Congress (WAC '94),* Maui, USA, August 1994, vol. 2, pp. 359–365.

[26] M. A. Unseren, "A New Technique for Dynamic Load Distribution when Two Manipulators Mutually Lift a Rigid Object. Part 2: Derivation of Entire System Model and Control Architecture," in *Proc. First World Automation Congress (WAC '94),* Maui, USA, August 1994, vol. 2, pp. 367–372.

[27] M. Uchiyama and T. Yamashita, "Adaptive Load Sharing for Hybrid Controlled Two Cooperative Manipulators," in *Proc. 1991 IEEE Int. Conf. on Robotics and Automation,* Sacramento, USA, April 1991, pp. 986–991.

[28] M. Uchiyama and Y. Kanamori, "Quadratic Programming for Dextrous Dual-Arm Manipulation," in *Robotics, Mechatronics and Manufacturing Systems, Trans. IMACS/SICE Int. Symp. on Robotics, Mechatronics and Manufacturing Systems, Kobe, Japan, September 1992*, Elsevier Science Publishers B.V. (North-Holland), pp. 367–372, 1993.

[29] M. Uchiyama, X. Delebarre, H. Amada, and T. Kitano, "Optimum Internal Force Control for Two Cooperative Robots to Carry an Object," in *Proc. First World Automation Congress (WAC '94)*, Maui, USA, August 1994, vol. 2, pp. 111–116.

[30] H. Inoue, "Computer Controlled Bilateral Manipulator," *Bul. JSME*, vol. 14, no. 69, pp. 199–207, 1971.

[31] M. Uchiyama, T. Kitano, Y. Tanno, and K. Miyawaki, "Cooperative Multiple Robots to Be Applied to Industries," in *Proc. World Automation Congress (WAC '96)*, Montpellier, France, May 1996, vol. 3, pp. 759–764.

[32] Y. F. Zheng and M. Z. Chen, "Trajectory Planning for Two Manipulators to Deform Flexible Beams," in *Proc. 1993 IEEE Int. Conf. on Robotics and Automation*, Atlanta, USA, May 1993, pp. 1019–1024.

[33] M. M. Svinin and M. Uchiyama, "Coordinated Dynamic Control of a System of Manipulators Coupled via a Flexible Object," in *Prepr. 4th IFAC Symp. on Robot Control*, Capri, Italy, September 1994, pp. 1005–1010.

[34] T. Yukawa, M. Uchiyama, D. N. Nenchev, and H. Inooka, "Stability of Control System in Handling of a Flexible Object by Rigid Arm Robots," in *Proc. 1996 IEEE Int. Conf. on Robotics and Automation*, Minneapolis, USA, April 1996, pp. 2332–2339.

[35] D. Sun, X. Shi, and Y. Liu, "Modeling and Cooperation of Two-Arm Robotic System Manipulating a Deformable Object," in *Proc. 1996 IEEE Int. Conf. on Robotics and Automation*, Minneapolis, USA, April 1996, pp. 2346–2351.

[36] J.-S. Kim, M. Yamano, and M. Uchiyama, "Lumped-Parameter Modeling for Cooperative Control of Two Flexible Manipulators," *Asia-Pacific Vibration Conf. '97*, Kyongju, Korea, November 1997, (to be presented).

[37] M. Yamano, J.-S. Kim, and M. Uchiyama, "Experiments on Cooperative Control of Two Flexible Manipulators Working in 3D Space," *Asia-Pacific Vibration Conf. '97*, Kyongju, Korea, November 1997, (to be presented).

[38] M. Uchiyama and A. Konno, "Modeling, Controllability and Vibration Suppression of 3D Flexible Robots," *Robotics Research, The Seventh International Symposium*, G. Giralt and G. Hirzinger (Eds), Springer, pp. 90–99, 1996.

[39] K. Munawar and M. Uchiyama, "Slip Compensated Manipulation with Cooperating Multiple Robots," *36th IEEE CDC*, San Diego, USA, December 1997, (to be presented).

[40] Y. Q. Dai, A. A. Loukianov, and M. Uchiyama, "A Hybrid Numerical Method for Solving the Inverse Kinematics of a Class of Spatial Flexible Manipulators," in *Proc. 1997 IEEE Int. Conf. on Robotics and Automation*, Albuquerque, USA, April 1997, pp. 3449–3454.

[41] M. Uchiyama, A. Konno, T. Uchiyama, and S. Kanda, "Development of a Flexible Dual-Arm Manipulator Testbed for Space Robotics," in *Proc. IEEE Int. Workshop on Intelligent Robots and Systems*, Tsuchiura, Japan, July 1990, pp. 375–381.

[42] P. Chiacchio, S. Chiaverini, and B. Siciliano, "Direct and inverse kinematics for coordinated motion tasks of a two-manipulator system," *Trans. ASME J. of Dynamic Systems, Measurement, and Control*, vol. 118, pp. 691–697, 1996.

[43] K. Kosuge and T. Oosumi, "Decentralized Control of Multiple Robots Handling an Object," in *Proc. 1996 IEEE/RSJ Int. Conf. on Intelligent Robots and Systems*, Osaka, Japan, November 1996, pp. 318–323.

Chapter 2

Kinematic manipulability of general mechanical systems

This paper extends the kinematic manipulability concept commonly used for serial manipulators to general constrained rigid multibody systems. Examples of such systems include multiple cooperating manipulators, multiple fingers holding a payload, multi-leg walking robots, and variable geometry trusses. Explicit formulas for velocity and force manipulability ellipsoids are derived and their duality explained. The concept of unstable grasp and manipulable grasp are also extended and illustrated with examples. It is then shown that manipulability can be significantly modified through bracing with another arm. Finally, several methods for comparing manipulability ellipsoids are developed which can be used in turn to optimize the brace design.

2.1   Introduction

This paper considers the kinematic manipulability of general constrained multibody systems. Such systems include a single articulated robot in contact with the environment, a multi-finger hand (Figure 2.1), multiple cooperative robots, and even a Stewart Platform (Figure 2.2). We first present a general kinematic model which considers all degrees of freedom and then imposes the constraints as algebraic conditions. Kinematic models of multi-finger grasping and a 6-DOF Stewart Platform are used as illus-

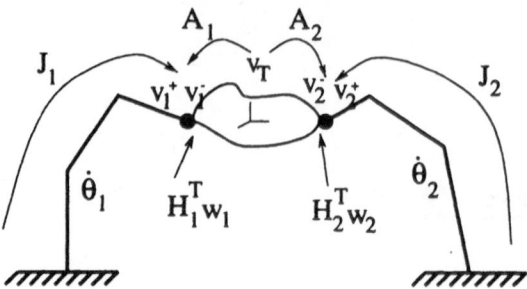

Figure 2.1: Two constrained manipulators in a load-sharing configuration.

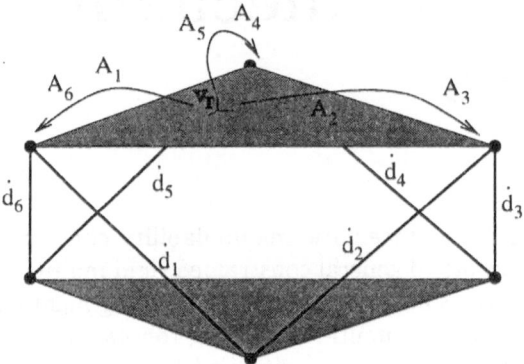

Figure 2.2: A Stewart platform.

trative examples. Through the Principle of Virtual Force, we also derive the general static force balance model which can be considered as a dual of the differential kinematics.

We then extend the familiar single arm manipulability ellipsoid concept first proposed in [1]. Characterization for both velocity and force ellipsoids is presented. When applied to multiple cooperative arms employing a rigid grasp or to multiple finger grasping, this work is closest to the work by [2] and is also closely related to the past work by [3, 4]. We also extend the important concepts of grasp stability and manipulability. We obtain explicit characterization for both properties and present their physical interpretation. As illustrations, we include a planar Stewart Platform, a full 6-DOF Stewart Platform, and a planar two-finger grasping example from [3, 4].

We also consider the effect of bracing on manipulability, by using a manipulator as a mobile fixture. Finding the best location and grasp type for a fixturing manipulator may be posed as a kinematic optimization problem, but a metric must first be defined for the manipulability ellipsoid. Several possible choices for ellipsoids metrics are stated and compared.

This paper is laid out in the following manner. We will first present the differential kinematic and static force model of a general constrained multiple-manipulator systems in Section 2.2. The velocity and force ellipsoids, and extension of grasp stability and manipulability are presented in Section 2.3. Section 2.4 presents a number of examples.

Terminology and Notation: We shall use the term "spatial force" at a given frame to mean the 6×1 vector of $\begin{bmatrix} \text{torque} \\ \text{force} \end{bmatrix}$, and the term "spatial velocity" at a given frame to mean the 6×1 vector $\begin{bmatrix} \text{angular velocity} \\ \text{linear velocity} \end{bmatrix}$. Given a matrix G, we use $\widetilde{G}$ to either denote the annihilator of G ($\widetilde{G}G = 0$) or the transpose of the annihilator of G^T ($G\widetilde{G} = 0$). Which of the two cases will be clear from the context.

2.2 Differential kinematics and static force model

This section considers the differential kinematics and static force balance of general rigid multibody systems. Multiple-finger grasping and a Stewart Platform will be used as examples.

2.2.1 Differential kinematics

We consider a general mechanism subject to kinematic constraints. The generalized coordinate (with the constraints removed) is denoted by θ. The active joints' angles are denoted by θ_a and passive ones by θ_p. We order the angles so that $\theta^T = [\theta_a^T, \theta_p^T]$. Consider a general constraint (written in terms of the joint velocity vector)

$$J_C(\theta)\dot{\theta} = 0. \tag{2.1}$$

Let the spatial velocity of the task frame be

$$v_T = J_T(\theta)\dot{\theta}. \tag{2.2}$$

Suppose that $J_C(\theta)$ is full rank. Then $\dot{\theta} = \tilde{J}_C\xi$, where $\tilde{J}_C^T$ is the annihilator of J_C^T. The task velocity can be written as

$$v_T = J_T\tilde{J}_C\xi. \tag{2.3}$$

The mechanism is singular if $J_T\tilde{J}_C$ loses rank; in other words, there are some directions in v_T that cannot be attained (but which can be attained in other arm configurations).

As an example, consider the kinematic model of multiple fingers grasping a rigid payload. For each serial chain, the joint velocity vector is defined as $\dot{\theta}_i$, the arm tip spatial velocity is v_i^+, and they are related by the arm Jacobian $J_i(\theta_i)$:

$$v_i^+ = J_i(\theta_i)\dot{\theta}_i$$

We consider a single task frame attached to the constraining rigid body (see Figure 2.1) whose spatial velocity is defined as v_T. On the payload side of the contact, the spatial velocity, v_i^-, is related to the task velocity v_T by

$$v_i^- = A_i v_T \quad \text{where} \quad A_i = \begin{bmatrix} I & 0 \\ L_{it}\times & I \end{bmatrix} \tag{2.4}$$

where L_{it} is the vector from the ith tip to the task frame. The relative velocity at each contact is parameterized by a velocity vector W_i:

$$v_i^+ + H_i^T W_i = v_i^-$$

where the columns of H_i^T are the directions where relative velocities at the contact are allowed.

To write the multi-arm kinematics more compactly, we stack all the vectors (e.g., θ_i's are stacked into a single vector θ_a) and block diagonalize all the matrices (e.g., J_i's form the diagonal blocks of J), except for $A^T \triangleq [\ A_1^T \ \dots \ A_m^T\ ]$. Then the differential kinematic relationship can be written as

$$
\begin{aligned}
v^+ + H^T W &= v^- \\
v^+ &= J\dot{\theta}_a \\
v^- &= A v_T
\end{aligned}
\tag{2.5}
$$

Some examples of possible contacts are shown in Figures 2.3–2.6.

Defining $\dot{\theta}_p = W$, we can represent the multi-finger kinematic model in the general form as in (2.1)-(2.2):

$$\underbrace{\tilde{A}\,[\ J \quad H^T\ ]}_{J_C(\theta)}\begin{bmatrix} \dot{\theta}_a \\ \dot{\theta}_p \end{bmatrix} = 0 \tag{2.6}$$

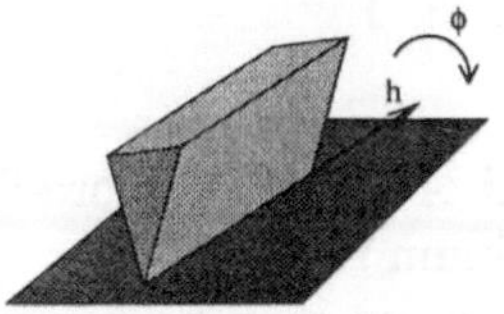

$$H^T W = \begin{bmatrix} h \\ 0 \end{bmatrix} \dot{\phi}$$

(6x1)   (6x1)(1x1)

Figure 2.3: line contact (1 DOF — rotation about h permitted).

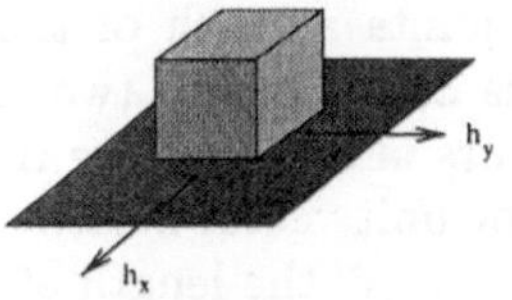

$$H^T W = \begin{bmatrix} 0 & 0 \\ h_x & h_y \end{bmatrix} \begin{bmatrix} v_x \\ v_y \end{bmatrix}$$

(6x1)   (6x2)   (2x1)

Figure 2.4: Sliding contact (2 DOF — sliding along h_x and h_y permitted).

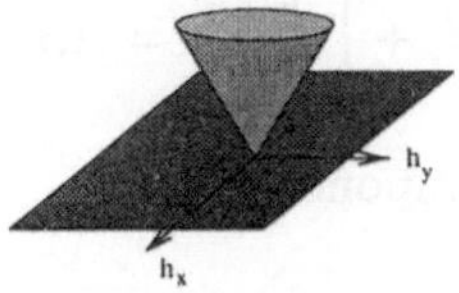

$$H^T W = \begin{bmatrix} I & 0 & 0 \\ 0 & h_x & h_y \end{bmatrix} \begin{bmatrix} \omega \\ v_x \\ v_y \end{bmatrix}$$

(6x1)   (6x5)   (5x1)

Figure 2.5: Point contact (5 DOF — rotation, sliding along h_x and h_y permitted).

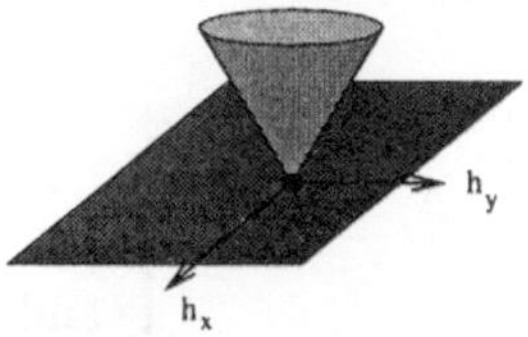

$$H^T W = \begin{bmatrix} I \\ 0 \end{bmatrix} \omega$$

(6x1)   (6x3)(3x1)

Figure 2.6: Point contact with friction (3 DOF — only rotation permitted).

$$\underbrace{\tilde{A}^+ \begin{bmatrix} J & H^T \end{bmatrix}}_{J_T(\theta)} \begin{bmatrix} \dot{\theta}_a \\ \dot{\theta}_p \end{bmatrix} = v_T \qquad (2.7)$$

where $\tilde{A}$ is the annihilator for A and A^+ is the Moore-Penrose pseudo-inverse of A. Note that A is of full column rank.

It is important to note that information may be removed from J and H^T prior to calculating $J_T(\theta)$. For example, if orientation is not important for the task to be performed, it may be useful to remove the orientation components of J and H^T, and calculate a simpler form for J_T. However, the constraint Jacobian J_C should contain full information about the system.

For another example, consider a Stewart Platform which consists of two triangular plates, with spherical joints at each of their three nodes (see Figure 2.2). Each bottom node is connected to two top nodes via a linear actuator, so there are six actuators in all. Suppose the task frame is attached rigidly to the top plate. Let the unit vector attached to each linear actuator be denoted by e_i, where $i = 1, \ldots, 6$, the length of the connection be d_i, the angular velocity of each leg be ω_i, and the angular velocity between the top plate and leg i be W_i. The rigid body transformation between the task frame and the top node connected to the ith leg is denoted by A_i (as given in (2.4). The kinematics then becomes

$$\begin{bmatrix} \omega_i \\ -d_i e_i \times \omega_i \end{bmatrix} + \begin{bmatrix} 0 \\ e_i \end{bmatrix} \dot{d}_i + \begin{bmatrix} W_i \\ 0 \end{bmatrix} = A_i v_T$$

Define a joint velocity vector with 42 components:

$$\dot{\theta} = \begin{bmatrix} \dot{d}_1 & \ldots & \dot{d}_6 & \omega_1 & W_1 & \ldots & \omega_6 & W_6 \end{bmatrix}^T. \qquad (2.8)$$

Note that $\dot{d}_1$ to $\dot{d}_6$ are active and others are passive. Stacking all the kinematic relations up vectorially, we have

$$J\dot{\theta} = A v_T \qquad (2.9)$$

where

$$J = \begin{bmatrix} 0 & & I & I & & \\ e_1 & & -d_1 e_1 \times & 0 & & \\ & \ddots & & & \ddots & \\ & & 0 & & I & I \\ & & e_6 & & -d_6 e_6 \times & 0 \end{bmatrix} \quad \text{and} \quad A = \begin{bmatrix} A_1 \\ \vdots \\ A_6 \end{bmatrix}$$

$$(2.10)$$

Eq. (2.9) can be equivalently written as

$$A^+ J\dot{\theta} = v_T$$
$$\tilde{A} J\dot{\theta} = 0.$$

In addition, the legs are constrain d so they cannot spin about themselves, so

$$e_i^T \omega_i = 0$$

which can also be written in terms of $\dot{\theta}$ as $J_{c1}\dot{\theta} = 0$ where J_{c1} is 6×42. Putting the constraints together, we have J_C in (2.1) as

$$J_C = \begin{bmatrix} J_{c1} \\ \tilde{A}J \end{bmatrix}. \tag{2.11}$$

2.2.2 Force balance

Static force balance can be considered as a dual to the kinematics. However, there is also the additional complication of static load such as gravity on each link and position feedback on the joint torque. We assume that these loads have already been excluded from the joint torque, or more specifically, we consider the joint torque τ to be the portion that balances with the load torque f_T (the force that the arm exerts at frame T). In the serial arm case, the force balance is simply $\tau = J^T(\theta) f_T$, where τ is the joint torque. This follows from the Principle of Virtual Work:

$$\tau^T \dot{\theta} = f_T^T v_T = f_T^T J(\theta)\dot{\theta}.$$

Since this holds true for any $\dot{\theta}$, the stated force relationship follows.

In the constrained mechanism case, we can apply the Principle of Virtual Work in a similar fashion (using the differential kinematic relationship (2.2)-(2.3) and noting that τ is now applied only at the active joints):

$$\begin{bmatrix} \tau \\ 0 \end{bmatrix}^T \dot{\theta} = \begin{bmatrix} \tau \\ 0 \end{bmatrix}^T \tilde{J}_C \xi = f_T^T v_T = f_T^T J_T \dot{\theta} = f_T^T J_T \tilde{J}_C \xi.$$

Since this holds true for any ξ, we have the force balance equation:

$$\tilde{J}_C^T \begin{bmatrix} \tau \\ 0 \end{bmatrix} = \tilde{J}_C^T J_T^T f_T. \tag{2.12}$$

This can be equivalently stated as

$$\begin{bmatrix} \tau \\ 0 \end{bmatrix} = J_T^T f_T + J_C^T \eta_T \tag{2.13}$$

where η_T is the "internal force" (in the multiple-arm context, the squeeze force).

The above can be viewed from another perspective. Instead of the constraint (2.1), we replace it with a "virtual velocity" (in the same spirit as in [5] in the multiple-arm rigid grasp context):

$$v_C = J_C \dot{\theta}. \tag{2.14}$$

Applying the Principle of Virtual Work again, we obtain

$$\begin{bmatrix} \tau \\ 0 \end{bmatrix}^T \dot{\theta} = f_T^T v_T + f_C^T v_C = (f_T^T J_T + f_C^T J_C)\dot{\theta} \tag{2.15}$$

where f_c is the force that enforces the constraint (2.1). Since the explicit constraint is removed, we have

$$\begin{bmatrix} \tau \\ 0 \end{bmatrix} = J_T^T f_T + J_C^T f_C. \tag{2.16}$$

This shows that the internal force η_T in (2.13) is actually the force that enforces the constraint (2.1).

As an aside, it should be noted that in mechanism design, it is important to know the internal loading, f_C, for a given amount of actuator torque, τ, and task loading, f_T. This can be done unambiguously if $\mathcal{N}(J_C^T) = \{0\}$ (where $\mathcal{N}(\cdot)$ denotes the null space). Equivalently, this means that the total number of unconstrained degrees of freedom (dimension of $\dot{\theta}$) is at least as many as the number of independent constraints. Otherwise, one has an underdetermined problem for the constraint force. This problem has been noted in the walking robot literature [6, 7].

We now apply the general frame work to the specific example of multi-finger grasping. The force relationship is given by

$$\tau = J^T f \qquad H f = 0 \qquad f_T = A^T f \tag{2.17}$$

which states that the stacked contact force f is zero in the direction where the contact is unconstrained (i.e., where relative motion is allowed) and the contact forces sum at the task frame to f_T. Solving f in terms of f_T, we have:

$$f = (A^T)^+ f_T + \tilde{A}^T f_C$$

where f_C is the force that enforces the constraint. Substituting into the τ equation and the contact constraint equation, we obtain (2.13):

$$\begin{bmatrix} \tau \\ 0 \end{bmatrix} = \underbrace{\begin{bmatrix} J^T \\ H \end{bmatrix} (A^T)^+}_{J_T^T} f_T + \underbrace{\begin{bmatrix} J^T \\ H \end{bmatrix} \tilde{A}^T}_{J_C^T} f_C. \tag{2.18}$$

As a specific example, consider two fingers pressing against each other with a frictional point contact. In the absence of the load force, f_T, we have the force balance

$$\begin{bmatrix} \tau_1 \\ \tau_2 \\ 0 \\ 0 \end{bmatrix} = \begin{bmatrix} J_1^T \\ -J_2^T \\ [I\,,\,0] \\ [-I\,,\,0] \end{bmatrix} f_C.$$

The last two sets of equations mean that f_C is a pure force (no torque component). The first two equations mean that the force due to the first finger is exactly balanced with the force from the second finger.

2.3 Velocity and force manipulability ellipsoids

2.3.1 Serial manipulators

The velocity manipulability ellipsoid of a single, serially-linked manipulator was introduced in [1] as an indication of the relative capability of a robot arm to move in different directions. Singular value decomposition (SVD) of the Jacobian, J, is the key tool in this analysis:

$$J = U\Sigma V^T \tag{2.19}$$

where U and V are orthogonal matrices, and Σ consists of a diagonal matrix with rows or columns of zeros added so that its dimension is the same as that of J. The Jacobian maps a ball in the joint velocity space to an ellipsoid in the spatial task velocity space:

$$\mathcal{E}_V = \left\{ v_T \,:\, v_T = J\dot{\theta}, \left\| \dot{\theta} \right\| = 1 \right\}.$$

The principal axes of the ellipsoid are given by the columns of U (left singular vectors), u_i's, and the lengths are given by the singular values, σ_i's. The right singular vectors, v_i's, (v_i^T is the ith row of V) are the preimage of u_i's: $Jv_i = \sigma_i u_i$. If J is less than full rank, then one or more principal axes of the ellipsoid will have zero length, and the ellipsoid will have zero volume. We say that the ellipsoid is *degenerate* in this case. If the ellipsoid is degenerate for all configurations (for example, for an arm with less than 6 DOF), then we can restrict the spatial task velocity to a lower dimensional manifold so that the ellipsoid is not degenerate at least for some configurations. If the rank of the Jacobian drops below its maximum rank at certain configurations, the arm is said to be *singular* in those configurations. With the spatial task velocity suitably restricted, singular configurations would correspond to degenerate ellipsoids. In this paper, we shall always assume that the maximum *row rank* of J over all possible configurations is full (i.e., $\mathcal{N}(J^T) = \{0\}$); this necessarily means that J is square or fat (redundant arm). Otherwise, the range of J can be suitably restricted (for all configurations) so this assumption would satisfy.

As a dual to the velocity ellipsoid, the force ellipsoid has also been introduced in the literature as the image in the end effector force space corresponding to a ball in the joint torque space:

$$\mathcal{E}_F = \left\{ f_T \; : \; J^T f_T = \tau, \|\tau\| = 1 \right\}.$$

By applying the SVD to J, we have $V\Sigma^T U^T f_T = \tau$. The non-degeneracy assumption means that $\Sigma^T = \begin{bmatrix} \Sigma_1 \\ 0 \end{bmatrix}$ where Σ_1 is square, diagonal, and full rank for at least some configurations. Partition $V = \begin{bmatrix} V_1 & V_2 \end{bmatrix}$ with dimensions compatible with Σ_1. Then

$$\begin{bmatrix} \Sigma_1 U^T f_T \\ 0 \end{bmatrix} = \begin{bmatrix} V_1^T \tau \\ V_2^T \tau \end{bmatrix}.$$

The bottom half of the above says that certain combination of joint torques cancel one another and does not produce an effector spatial force. They correspond to the self motion of a redundant arm. Solving the top half we obtain:

$$\mathcal{E}_F = \left\{ f_T \; : \; f_T = U\Sigma_1^{-1} V_1^T \tau, \|\tau\| = 1 \right\}.$$

This means that the principal axes of the force ellipsoid are the same as the velocity ellipsoid, but the lengths are the reciprocal of those in the velocity ellipsoid. When the arm is in a singular configuration, the null space of J^T would be non-zero (or one or more diagonal entries in Σ_1 are zero), implying that the force ellipsoid is infinite in the corresponding directions in U. Such configurations restrict motion but are mechanically advantageous as the mechanism can (theoretically) bear infinite load in certain direction.

In this section, we present an extension of these concepts to general constrained mechanisms. For the specific cases of multi-finger grasp, the development here is similar to that in [3, 4] and the more recent work in [2].

2.3.2   Velocity ellipsoid

Consider the general kinematic equation (2.1)-(2.2). The unconstrained Jacobian, J_T, maps a unit ball in the joint velocity space to an ellipsoid in the tip contact velocity space. Due to the constraint (2.1), only a certain slice of the ball (resp., ellipsoid) is feasible. It is reasonable to define the constrained ellipsoid as the set of spatial task velocities generated by a unit ball in the *active* joint velocity space:

$$\mathcal{E}_V = \left\{ v_T : \left\| \dot{\theta}_a \right\| = 1, v_T = J_T\dot{\theta}, J_C\dot{\theta} = 0 \right\} \qquad (2.20)$$

Substituting the parameterization as in (2.3) and partitioning J_C and $\tilde{J}_C$ (corresponding to the active and passive joints, respectively) as

$$J_C = \begin{bmatrix} J_{c_a} & J_{c_p} \end{bmatrix} \qquad \tilde{J}_C = \begin{bmatrix} \tilde{J}_{c_a} \\ \tilde{J}_{c_p} \end{bmatrix}$$

then the constrained ellipsoid can be written as

$$\mathcal{E}_V = \left\{ v_T : v_T = J_T\tilde{J}_C\xi, \left\| \tilde{J}_{c_a}\xi \right\| = 1 \right\}. \qquad (2.21)$$

We shall consider three cases:

1. *No independent passive joint motion* $\mathcal{N}(\tilde{J}_{c_a}) = \{0\}$. This means that if the active joints are locked, the entire mechanism is also locked. An example of this case is a stable multi-finger grasp.

2. *No unactuated task motion* $\mathcal{N}(\tilde{J}_{c_a}) \neq \{0\}$ and $\mathcal{N}(\tilde{J}_{c_a}) \subset \mathcal{N}(J_T\tilde{J}_C)$. This means that there can be independent passive joint motion, but it does not produce any task motion. As an example, consider a Stewart Platform with all spherical joints at the nodes. Then each leg can spin about its own axis without causing motion of the task frame attached to the upper platform.

3. *Unactuated task motion* $\mathcal{N}(\tilde{J}_{c_a}) \neq \{0\}$ and $\mathcal{N}(\tilde{J}_{c_a}) \not\subset \mathcal{N}(J_T\tilde{J}_C)$. This case covers the remaining scenario: even if all the active joints are locked, there can still be task motion involving the passive joints. An unstable multi-finger grasp is an example of this case.

In the first two cases, the manipulability ellipsoid is still well defined. In the last case, the mechanism is in a sense *unstable*, and the manipulability ellipsoid would be infinite. Note that there is no counterpart to this case in the serial arm case. Even in the multi-finger literature, unstable grasp is rarely addressed — they are usually eliminated by assumption. We now address the above three cases in greater details.

Case 1. $\mathcal{N}(\widetilde{J}_{c_a}) = \{0\}$. The ellipsoid can be rewritten as

$$\mathcal{E}_V = \left\{ v_T : v_T = J_T \widetilde{J}_C \left(\widetilde{J}_{c_a}^T \widetilde{J}_{c_a} \right)^{-\frac{1}{2}} x, \|x\| = 1 \right\}. \qquad (2.22)$$

As in the unconstrained arm case, the singular values and left singular vectors of the reduced Jacobian $J_T \widetilde{J}_C \left(\widetilde{J}_{c_a}^T \widetilde{J}_{c_a} \right)^{-\frac{1}{2}}$ correspond to the length and direction of the principal axes of the multiple arm ellipsoid. It is also straightforward to include weighted norms in the joint and/or task spaces in the above definition.

Case 2. $\mathcal{N}(\widetilde{J}_{c_a}) \neq \{0\}$ and

$$\mathcal{N}(\widetilde{J}_{c_a}) \subset \mathcal{N}(J_T \widetilde{J}_C) \qquad (2.23)$$

In this case, the ellipsoid can be computed by removing the $\mathcal{N}(\widetilde{J}_{c_a})$ component in (2.21). To this end, let $K = \begin{bmatrix} K_1 & K_2 \end{bmatrix}$ where sp$\{$ col $(K_1) \} = \mathcal{R}(\widetilde{J}_{c_a}^T)$ and sp$\{$ col $(K_2) \} = \mathcal{N}(\widetilde{J}_{c_a})$. By construction, K is square invertible. Then under the assumption (2.23),

$$\begin{aligned} \mathcal{E}_V &= \left\{ v_T : v_T = J_T \widetilde{J}_C [K_1 \; 0] K^{-1} \xi; \; \left\| \widetilde{J}_{c_a} [K_1 \; 0] K^{-1} \xi \right\| = 1 \right\} \\ &= \left\{ v_T : v_T = J_T \widetilde{J}_C K_1 [(\widetilde{J}_{c_a} K_1)^T (K_1 \widetilde{J}_{c_a})]^{-\frac{1}{2}} x, \|x\| = 1 \right\} \end{aligned}$$

The second equality is obtained by eliminating the bottom portion of $K^{-1}\xi$. The ellipsoid can be computed from SVD of $J_T \widetilde{J}_C K_1$ $[(\widetilde{J}_{c_a} K_1)^T (\widetilde{J}_{c_a} K_1)]^{-\frac{1}{2}}$. Note that by construction, $\mathcal{N}(\widetilde{J}_{c_a} K_1) = \{0\}$.

Case 3. $\mathcal{N}(\widetilde{J}_{c_a}) \neq \{0\}$ and

$$\mathcal{N}(\widetilde{J}_{c_a}) \not\subset \mathcal{N}(J_T \widetilde{J}_C). \qquad (2.24)$$

In this case, there exist $\xi \in \mathcal{N}(\widetilde{J}_{c_a})$ such that $\dot{\theta}_a = 0$ and $v_T \neq 0$, implying that the ellipsoid would be infinite in these directions. Such configurations are in a sense unstable (see the force ellipsoid section below for further discussion) and should be avoided. If such a situation is encountered, it may be tempting to consider the ellipsoid resulting from the motion of the active joints only. This ellipsoid is not meaningful since, for the same active joint velocity, there may be multiple possible task velocities, depending on the motion of the passive joints.

Manipulability ellipsoids also provide a geometric visualization for singular configurations. Suppose that the ellipsoid is not always degenerate (where the lengths of one or more axes become zero, implying that the ellipsoid has zero volume). Then the configurations at which the ellipsoid does become degenerate are the singular configurations. They can be found by solving for the zeros of the singular values of the Jacobian matrices discussed above.

2.3.3 Force ellipsoid

The force ellipsoid can be intuitively defined as the set of task forces that can be applied by the mechanism with active torques (or forces) constrained on the surface of a weighted ball. Recalling the constraint force balance equation (2.12), we obtain the dual of (2.21)

$$\mathcal{E}_F = \left\{ f_T : \tilde{J}_C^T J_T^T f_T = \tilde{J}_{c_a}^T \tau, \|\tau\| = 1 \right\}. \tag{2.25}$$

As in the single arm case, we assume that $\mathcal{N}(\tilde{J}_C^T J_T^T) = \{0\}$ except at singular configurations (i.e., the velocity manipulability ellipsoid is not always degenerate). If this is not satisfied, we can always suitably restrict f_T so it is true. Similar to the velocity ellipsoid case above, there are three cases to consider:

1. $\tilde{J}_{c_a}^T$ *is onto.* This condition means that the active joints can generate all forces corresponding to the independent degrees of freedom, ξ. Mathematically, this condition is also equivalent to the *Case 1* for the velocity ellipsoid, $\mathcal{N}(\tilde{J}_{c_a}) = \{0\}$.

2. $\tilde{J}_{c_a}^T$ *is not onto and*

$$\mathcal{R}(\tilde{J}_C^T J_T^T) \subset \mathcal{R}(\tilde{J}_{c_a}^T) \tag{2.26}$$

   In this case, active joints can generate all possible spatial forces in the task frame, but there are some internal forces (corresponding to motion) that cannot be generated. This condition is also equivalent to the *Case 2* for the velocity ellipsoid, $\mathcal{N}(\tilde{J}_{c_a}) \neq \{0\}$ and $\mathcal{N}(\tilde{J}_{c_a}) \subset \mathcal{N}(J_T \tilde{J}_C)$.

3. $\mathcal{R}(\tilde{J}_C^T J_T^T) \not\subset \mathcal{R}(\tilde{J}_{c_a}^T)$. For this remaining case, there are spatial task forces that cannot be generated by the active joint torques. The condition is also equivalent to the *Case 3* for the velocity ellipsoid, $\mathcal{N}(\tilde{J}_{c_a}) \not\subset \mathcal{N}(J_T \tilde{J}_C)$.

As in the single serial arm case, the ellipsoid computation is the dual of the velocity ellipsoid. We now elaborate each case below:

Case 1. Since $\tilde{J}_{c_a}$ is onto, the active joint torque τ can be decomposed as

$$\tau = \tilde{J}_{c_a}\eta_1 + J_{c_a}^T\eta_2.$$

It is clear that η_2 does not contribute to f_T and so can be ignored in the ellipsoid calculation. The force ellipsoid can then be written as:

$$\mathcal{E}_F = \left\{ f_T : (\tilde{J}_{c_a}^T\tilde{J}_{c_a})^{-1}(\tilde{J}_C^T J_T^T)f_T = \eta_1 \; \left\| \tilde{J}_{c_a}\eta_1 \right\| = 1 \right\}$$

$$= \left\{ f_T : (\tilde{J}_{c_a}^T\tilde{J}_{c_a})^{-\frac{1}{2}}(\tilde{J}_C^T J_T^T)f_T = y \; \|y\| = 1 \right\}.$$

Again as in the single serial arm case, if the SVD of the overall Jacobian is

$$J_T\tilde{J}_C \left(\tilde{J}_{c_a}^T\tilde{J}_{c_a} \right)^{-\frac{1}{2}} = U \left[\; \Sigma_1 \quad 0 \; \right] V^T$$

the force ellipsoid can be computed from $U\Sigma_1^{-1}V_1^T$.

Case 2. In this case, $\tilde{J}_{c_a}^T$ is no longer onto. We can recover the case above by projecting both sides of the force balance onto the range of $\tilde{J}_{c_a}^T$. Let $K = \left[\; K_1 \quad K_2 \; \right]$ be defined as in the previous section. Then

$$K^T J_T^T \tilde{J}_C^T f_T = K^T \tilde{J}_{c_a}^T \tau$$

$$\Longleftrightarrow \quad \left[\begin{array}{c} K_1^T J_T^T \tilde{J}_C^T f_T \\ 0 \end{array} \right] = \left[\begin{array}{c} K_1^T \tilde{J}_{c_a}^T \tau \\ 0 \end{array} \right].$$

The above equations means that any spatial force at the task frame would *only* affect the active joints and not the passive joints. Therefore, we only need to keep the top equation and obtain the dual of *Case 2* of the velocity ellipsoid. If the SVD of the overall Jacobian $J_T\tilde{J}_C K_1 \left[(\tilde{J}_{c_a}K_1)^T(\tilde{J}_{c_a}K_1) \right]^{-\frac{1}{2}}$ is $U \left[\; \Sigma_1 \quad 0 \; \right] V^T$, then the force ellipsoid can be computed from $U\Sigma_1^{-1}V_1^T$.

Case 3 As in *Case 2*, we can multiply K to both sides of the force balance again:

$$K^T J_T^T \tilde{J}_C^T f_T = K^T \tilde{J}_{c_a}^T \tau$$

$$\Longleftrightarrow \quad \left[\begin{array}{c} K_1^T J_T^T \tilde{J}_C^T f_T \\ K_2^T J_T^T \tilde{J}_C^T f_T \end{array} \right] = \left[\begin{array}{c} K_1^T \tilde{J}_{c_a}^T \tau \\ 0 \end{array} \right]. \tag{2.27}$$

This means that spatial force at the task frame not only will affect the active joints but will load the passive joints as well. Since the passive joints cannot resist such load, uncontrolled motion will result. The task frame forces that will load the passive joints are those in the range of $\tilde{J}_C J_T K_2$. To avoid uncontrolled motion, there can be no external load in this subspace. This condition (the bottom half of (2.27)) means that the force ellipsoid is a slice of the ellipsoid from the top half of (2.27). In other words, the ellipsoid is degenerate (or zero volume).

2.3.4   Configuration stability and manipulability

For multi-finger systems, there are two important concepts: grasp stability and grasp manipulability. A grasp is stable if any external force applied at the task frame can be resisted by suitably chosen joint torques. Equivalently, a grasp is also stable if there is no task motion independent from the joint motion. A classic example of an unstable grasp is two fingers holding a payload with frictional point contacts. The object can then spin about the line linking the contact points. Mathematically, the stable grasp condition can be stated as

$$\mathcal{N}(\tilde{H}^T A) = \{0\}.$$

where H^T and A are as defined in 2.5. A grasp is manipulable if any task velocity can be achieved with suitably chosen joint velocity. Mathematically, this condition can be stated as

$$\mathcal{R}(\tilde{H}^T J) \supset \mathcal{R}(\tilde{H}^T A).$$

where H^T, J, and A are as in 2.5

These concepts can be generalized to general constrained mechanisms. We will say that the mechanism is in a stable configuration if any external force applied at the task frame can be resisted by suitably chosen active joint force/torque, or equivalently, if there is no task motion independent from the active joint motion. Under this definition, it is clear that assumptions (2.23) or (2.26) is the condition for a stable configuration.

We can similarly define that a mechanism is manipulable if any task velocity can be achieved with suitably chosen active joint velocity. This simply means that the manipulability ellipsoid defined in the previous section is not degenerate (i.e., none of the principal axes has zero length). We have already made the assumption that the mechanism under consideration is manipulable except at singular configurations.

It is interesting to observe the dual relationship between unstable configurations and singular configurations. At a singular configuration, the velocity ellipsoid is degenerate (mechanism cannot move in certain directions) and the force ellipsoid is infinite (mechanism can resist infinite force in the same directions). At an unstable configuration, the force ellipsoid is degenerate (mechanism cannot resist force in certain directions) and the velocity ellipsoid is infinite (mechanism can have any velocity using only passive joints). In a near singular configuration, large joint motion may be required to achieve small task motion. Similarly, in a near unstable configuration, large joint torques may be required to counteract small external force applied at the task frame.

2.3.5 Internal force and virtual velocity

In (2.14), we introduced the concept of virtual velocity as the dual of the internal force. Similar to [8], we can also define a virtual velocity ellipsoid (resp. internal force ellipsoid) as the image of a unit ball of active joint velocity (resp. active joint torque) subject to the constraint that the spatial task velocity (resp. spatial task force) is zero:

$$\mathcal{E}_{V_C} = \left\{ v_C : v_C = J_C \dot{\theta}, J_T \dot{\theta} = 0, \left\| \dot{\theta}_a \right\| = 1 \right\} \tag{2.28}$$

$$\mathcal{E}_{F_C} = \left\{ f_C : \tilde{J}_T^T J_C^T f_C = \tilde{J}_T^T \begin{bmatrix} \tau \\ 0 \end{bmatrix}, \|\tau\| = 1 \right\}. \tag{2.29}$$

Mathematically, these ellipsoids are exactly the same as the velocity and force ellipsoids discussed before except that the subscripts T and C are exchanged. Therefore, all the preceding discussion on their computation remains valid. The concept of unstable configuration now translates to a degenerate internal force ellipsoid and infinite virtual velocity ellipsoid.

In a general mechanism, internal force may determine if a constraint can be enforced. For example, in a multi-finger grasp with frictional contacts, each contact force needs to be in the friction cone to ensure that the contact can be sustained. The internal force ellipsoid provides information on the ability that the active joints may impart on the internal force. Virtual velocity provides an appealing dual to the internal force, but it is not as practically significant.

2.4 Illustrative examples

2.4.1 Simple two-arm example

We first consider a planar two-finger grasping example. Figure 2.7 shows two two-joint fingers holding a rigid object (here depicted as a bar) between them. First consider the Jacobian for each arm mapping the joint angles to the tip *translational velocity*:

$$J_i = \begin{bmatrix} -\ell_{i1}s_{i1} - \ell_{i2}s_{i12} & -\ell_{i2}s_{i12} \\ \ell_{i1}c_{i1} + \ell_{i2}c_{i12} & \ell_{i2}c_{i12} \end{bmatrix} \begin{bmatrix} \dot{\theta}_{i1} \\ \dot{\theta}_{i2} \end{bmatrix}$$

where ℓ_{ij} is the length of the jth link of the ith arm, s_{ij} is the sine of the jth angle of the ith arm, s_{ijk} is the sine of the sum of the jth and kth angles of the ith arm. The task velocity (defined as the translational velocity of a specific point on the held bar) is related to the tip velocity as:

$$v_T = v_1 + \begin{bmatrix} -L_1s_1 \\ L_1c_1 \end{bmatrix} \dot{\theta}_1 = v_2 + \begin{bmatrix} -L_2s_2 \\ L_2c_2 \end{bmatrix} \dot{\theta}_2$$

where $s_i = sin(\theta_i)$, $c_i = cos(\theta_i)$, and θ_i denotes the angle at the ith contact. The overall kinematics is of the following form:

$$J\dot{\theta} + H^T W = A v_T \tag{2.30}$$

where $J = \text{diag}\{J_1, J_2\}$, $H^T = \text{diag}\{H_1^T, H_2^T\}$, and

$$W = \begin{bmatrix} \dot{\theta}_1 \\ \dot{\theta}_2 \end{bmatrix} \quad H_i^T = \begin{bmatrix} -L_i s_i \\ L_i c_i \end{bmatrix} \quad A = \begin{bmatrix} I \\ I \end{bmatrix}.$$

For the constraint, the orientation needs to be included. The corresponding kinematics are:

$$J^{(c)}\dot{\theta} + H^{(c)^T} W = A^{(c)} v_T \tag{2.31}$$

where

$$
\begin{aligned}
J^{(c)} &= \text{diag}\{J_1^{(c)}, J_2^{(c)}\} \\
J_i^{(c)} &= \begin{bmatrix} [1, 1] \\ J_i \end{bmatrix} \\
H^{(c)^T} &= \text{diag}\{H_1^{(c)^T}, H_2^{(c)^T}\} \\
H_i^{(c)^T} &= \begin{bmatrix} 1 \\ H_i^T \end{bmatrix} \\
A^{(c)} &= \begin{bmatrix} 0 & 0 \\ I_{2\times2} \\ 0 & 0 \\ I_{2\times2} \end{bmatrix}.
\end{aligned}
$$

The kinmeatics can be written as (2.1)-(2.2) with

$$J_C = \widetilde{A^{(c)}} \begin{bmatrix} J^{(c)} & H^{(c)^T} \end{bmatrix} \quad J_T = A^+ \begin{bmatrix} J & H^T \end{bmatrix}.$$

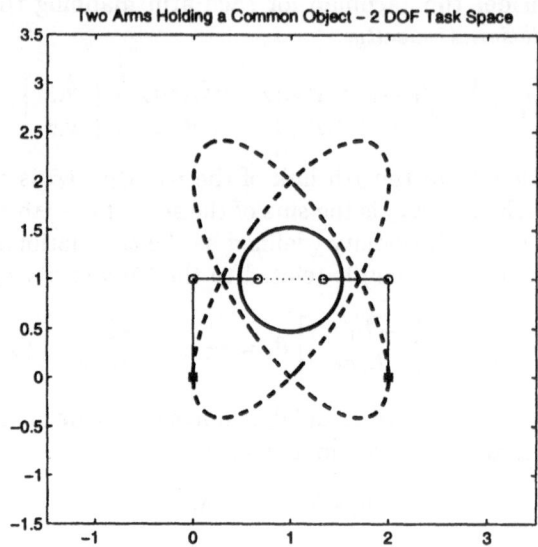

Figure 2.7: Two arms holding a rigid payload.

Consider in particular the configuration shown in Figure 2.7. Such an example was first suggested in [8], and discussed further in [9, 3, 10, 4]. The ellipsoid indicates that the system permits v_T in both the x and y directions. This makes sense since the robots are allowed to pivot at the contact points.

To prevent pivoting at the contact, we simply remove $H^T W$ in (2.30) and $H^{(c)^T} W$ in (2.31). In this case, the ellipsoid is degenerate and the task frame can only translate in the x direction. The degenerate ellipsoid (a horizontal line segment) is shown in Figure 2.8.

In [4], this example was used to demonstrate the superiority of the ellipsoid characterization as compared to those in [11, 8]. However, the key difference in terms of the nature of the grasp was not noted.

2.4.2 Planar Stewart platform example

We use a planar Stewart Platform to illustrate our approach applied to a mechanism that is *not* a single closed chain. Figure 2.9 shows various different planar Stewart Platforms each with 3 active prismatic joints and 6 passive rotational joints.

We consider the task velocity as the linear velocity of the center of the

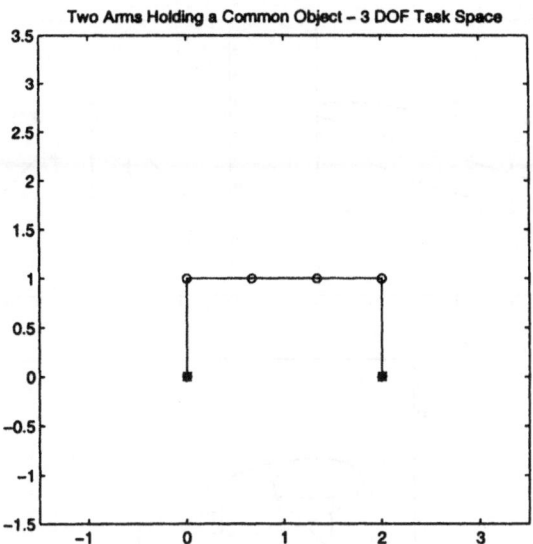

Figure 2.8: Manipulability ellipsoid with orientation consideration.

platform.

$$Av_T = J\dot{d} + H^T W \qquad (2.32)$$

where $J = \text{diag}\{J_1, J_2, J_3\}$, $H^T = \text{diag}\{H_1^T, H_2^T, H_3^T\}$, $d = [d_1, d_2, d_3]^T$ (prismatic joint velocities), $W = [W_1, W_2, W_3]^T$, $A = [I_{2\times2}, I_{2\times2}, I_{2\times2}]^T$

$$J_i = \begin{bmatrix} c_i \\ s_i \end{bmatrix}$$

$$W_i = \begin{bmatrix} \dot{\theta}_i \\ \dot{\phi}_i \end{bmatrix}$$

θ_i is the angular velocity of leg i with respect to the ground, ϕ_i is the angular velocity of the platform with respect to leg i, $c_i \triangleq cos(\theta_i)$, $s_i \triangleq sin(\theta_i)$.

As in the previous example, the constraint kinematics include orientation and therefore needs to be separately stated:

$$A^{(c)}v_T = J^{(c)}\dot{d} + H^{(c)^T} W \qquad (2.33)$$

where $J^{(c)} = \text{diag}\{J_1^{(c)}, J_2^{(c)}, J_3^{(c)}\}$, $H^{(c)^T} = \text{diag}\{H_1^{(c)^T}, H_2^{(c)^T}, H_3^{(c)^T}\}$, $A^{(c)^T} = [A_1^{(c)^T}, A_2^{(c)^T}, A_3^{(c)^T}]$ and

$$J_i^{(c)} = \begin{bmatrix} 0 \\ J_i \end{bmatrix}$$

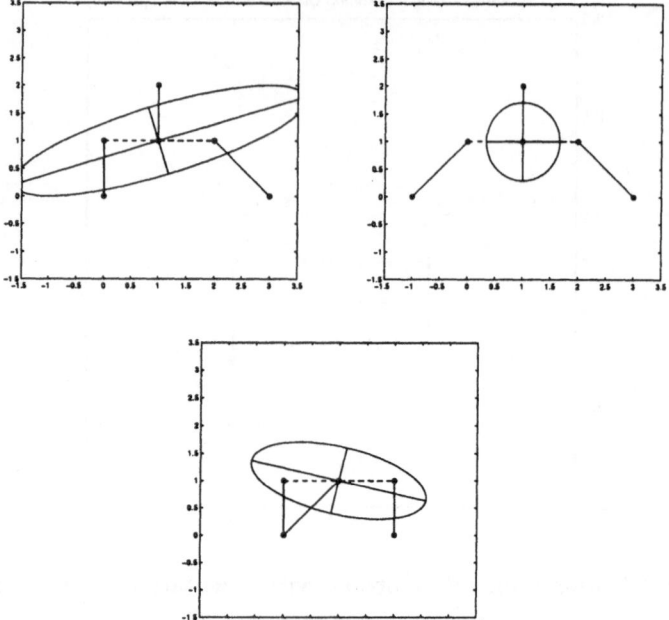

Figure 2.9: Velocity ellipsoids for various Stewart platforms.

$$H_i^{(c)^T} = \begin{bmatrix} [1,1] \\ H_i \end{bmatrix}$$

$$A_i^{(c)} = \begin{bmatrix} [0,0] \\ I_{2\times2} \end{bmatrix}.$$

Transforming these equations to the form that we have used, (2.1)-(2.2), we have

$$J_C = \widetilde{A^{(c)}} \begin{bmatrix} J^{(c)} & H^{(c)^T} \end{bmatrix} \tag{2.34}$$

$$J_T = A^+ \begin{bmatrix} J & H^T \end{bmatrix}. \tag{2.35}$$

Using the results presented earlier, ellipsoids for different configurations can be readily generated (as shown in Figure 2.9). All of these cases correspond to stable, nonsingular configurations.

For the configuration shown in Figure 2.10, $\mathcal{N}\{\tilde{J}_{c_a}\} \neq \{0\}$. For the case shown, the mechanism can have a pure horizontal motion involving only the passive joints ($\dot{\theta}_1 = \dot{\theta}_2 = -\dot{\theta}_3 = 1$). ¿From a force perspective, the

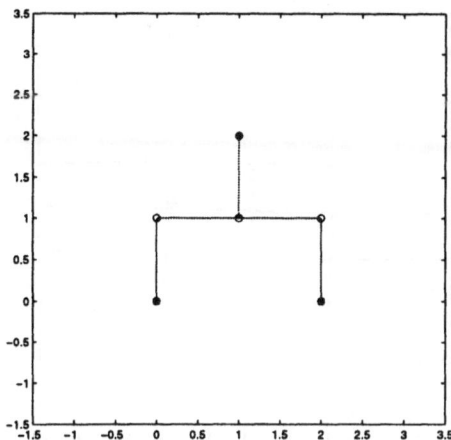

Figure 2.10: Unstable configuration.

unstable configuration means that the mechanism cannot resist x direction force applied at the task frame.

When the mechanism is near an unstable configuration, it may not be unstable mathematically, but the ellipsoid will be badly conditioned. As shown in Figure 2.11, the motion in the x direction is much larger than in the y direction. When the mechanism moves in to the unstable configuration, the ellipsoid becomes infinite in the x direction. From the force perspective, this suggests that a nearly unstable configuration is also highly undesirable as large forces from the active joints are needed to counteract disturbance force at the task frame. We have constructed a physical 3DOF Stewart Platform, and have indeed verified that unstable and nearly unstable configurations can have large internal motion with all the active joints locked. When the ellipsoid is well conditioned, such internal motion is no longer possible.

2.4.3 Six-DOF Stewart platform example

We now consider a 6-DOF Stewart Platform. Let the three base nodes be at

$$
x_1 = \begin{bmatrix} -1 \\ -1 \\ 0 \end{bmatrix} \quad x_2 = \begin{bmatrix} 1 \\ -1 \\ 0 \end{bmatrix} \quad x_3 = \begin{bmatrix} 0 \\ 1 \\ 0 \end{bmatrix}.
$$

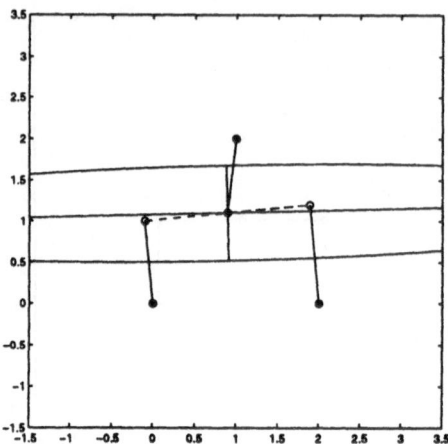

Figure 2.11: Nearly unstable configuration.

The top platform is an isosceles triangle with the two equal sides of length 1.12 and the third side of length 1. The task velocity, v_T, is defined as the translational velocity of the half way point of the line perpendicular to the base of the isosceles platform. As in the two previous examples, the task velocity only involves the linear motion but the constraints need to include orientation. Therefore, the kinematics developed in Section 2.2.1 needs to be slightly modified. With $\dot{\theta}$ as defined in (2.8), the task velocity kinematics is now

$$
\begin{bmatrix}
e_1 & & -d_1 e_1 \times & 0_{3\times 3} \\
& \ddots & & \ddots \\
& e_6 & & -d_6 e_6 \times & 0_{3\times 3}
\end{bmatrix}
\dot{\theta} =
\begin{bmatrix}
I_{3\times 3} \\
\vdots \\
I_{3\times 3}
\end{bmatrix}
v_T.
$$
(2.36)

The constraint equation, (2.1), is the same as in Section 2.2.1, given by (2.11).

The velocity ellipsoids of the Stewart Platform in three different configurations are shown in Figures 2.12– 2.14 (the force ellipsoids have the same principal axes but reciprocal length). In the first case, the platform is horizontal. In the second case, the task frame is rotated 45° about the axis $\begin{bmatrix} 0.71 & 0.71 & 0 \end{bmatrix}^T$. In the third case, the task frame is rotated 22.5° about the vertical axis $\begin{bmatrix} 0 & 0 & 1 \end{bmatrix}$.

In each case, three ellipses lying in the plane generated by two of the principal axes are shown. In the first case, the ellipse is well conditioned

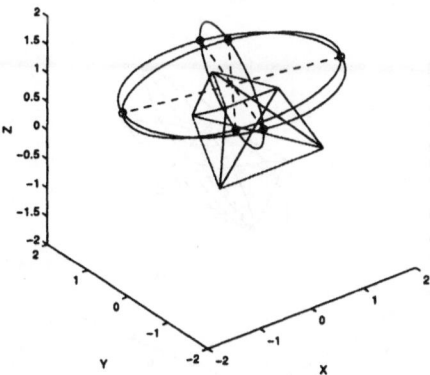

Figure 2.12: 3D ellipsoid for 6-DOF Stewart platforms: Case 1.

with the lengths of principal axes: $\{1.78, 1.43, 0.81\}$. In the second case, the ellipsoid becomes less well conditioned, the lengths of the principal axes are $\{2.31, 1.62, 0.29\}$. The motion parallel to the platform is more difficult than other directions. In the third case, the lengths of the principal axes are $\{5.62, 1.69, 1.49\}$. Even though the ellipsoid is fairly well conditioned (condition number of the singular values is 3.78), but external forces along the principal axis that corresponds to 5.62, $\begin{bmatrix} -0.54 & 0.12 & -0.83 \end{bmatrix}$, cannot be resisted as easily as in other directions.

2.5 Effects of arm posture and bracing on manipulability

In this section, we consider the effect of arm posture, bracing, and grasp type on the manipulability of the arm (and therefore the ellipsoid).

2.5.1 Effect of arm posture

For nonredundant arms, there is little choice in positioning the robot joints in order to allow the end-effector to perform some task. For redundant arms, there is much more flexibility, allowing the joints to be positioned in a way which makes it easier for the arm to perform the desired task.

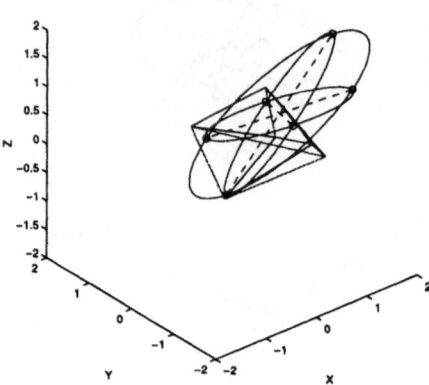

Figure 2.13: 3D ellipsoid for 6-DOF Stewart platforms: Case 2.

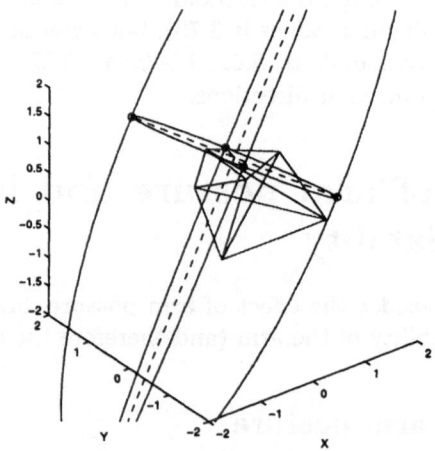

Figure 2.14: 3D ellipsoid for 6-DOF Stewart platforms: Case 3.

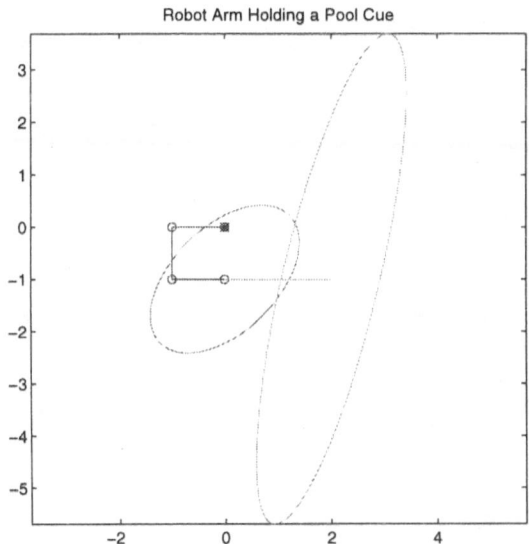

Figure 2.15: Ellipsoids for the end effector and for the tool tip.

An inefficient arm posture will require the motors to either apply more force to the joints in order to obtain some desired force at the end-effector, or to move the joints more quickly in order to achieve some desired end-effector velocity, than is necessary. If a change in the arm posture can improve the performance (efficiency) of the arm, it makes sense to alter the configuration of the robot.

Figure 2.15 shows a 3 DOF (redundant) planar robot arm, holding a pool cue straight out to the right. For simplicity, all robot links are of length 1, and the cue is of length 2. The arm is shown in red. The ellipsoid for the end-effector is shown in green, while the pool cue and the ellipsoid at the cue's tip is shown in light blue. The ellipsoids indicate the ability of the end-effector and the cue's end to move in the x or y directions (i.e. rotation is not considered).

Figures 2.16 and 2.17 show this same robot arm in a variety of different postures, and the manipulability ellipsoid at the tool tip in each case. In all of the figures, the location of the end effector is the same (1 unit below the base of the robot). From the figures, it is clear that the arm posture can have a major effect on the shape and orientation of the ellipsoid - and thus, its manipulability.

Applying the ellipsoid metrics here can provide more insight into the

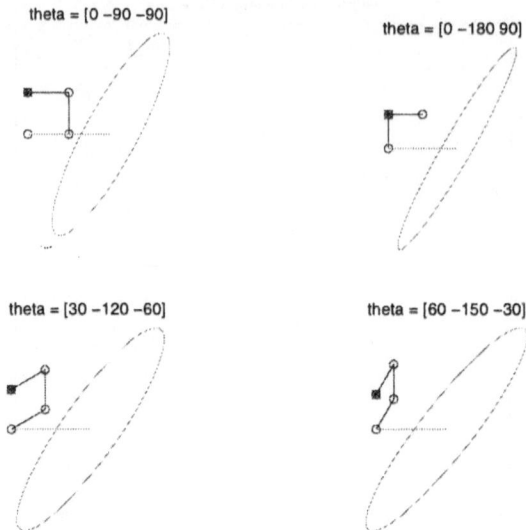

Figure 2.16: Effect of different arm postures on the manipulability ellipsoid.

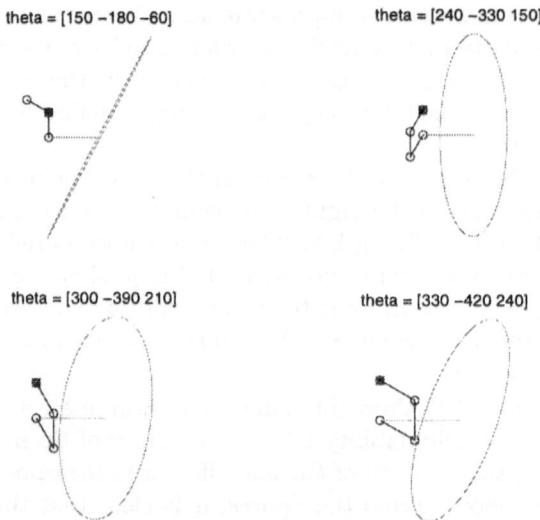

Figure 2.17: Effect of different arm postures on the manipulability ellipsoid.

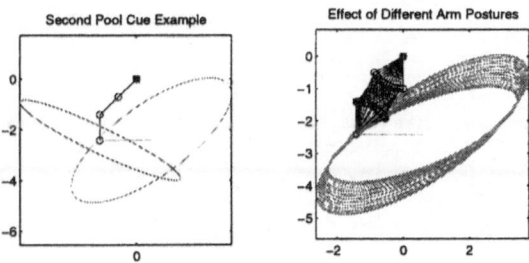

Figure 2.18: Effect of arm postures on the manipulability ellipsoid: Second example.

effect that the arm posture may have on the manipulability in this example. A comparison of a large number of the possible manipulability ellipsoids indicates that shape, scale and rotation of the ellipsoids are all affected by the arm posture. The largest distance between the various ellipsoids was found to be: $\alpha : 0.92$, $\beta : 0.44$, $\gamma : 2.17$, $\delta : 0$. Only translation has not been affected, since the end effector could always be placed in the same location.

Figures 2.18 and 2.19 show this same robot arm holding the tool at a different location. The manipulability ellipsoid for the end-effector is shown in green, while the ellipsoid for the tool tip is shown in light blue. The second part of figure 2.18 shows several arm configurations, and their ellipsoids all superimposed on each other; from this, one can get a feel for the how much the ellipsoid can be shaped by arm posture in this case. Figure 2.19 shows 4 different individual arm postures, with their corresponding ellipsoids.

The largest "distance" between the various ellipsoids was found to be: $\alpha : 0.33$, $\beta : 0.07$, $\gamma : 0.75$, $\delta : 0$. Note that all of the metric results are less than in the previous example. This indicates that the arm posture does not have has much effect on the shape of the ellipsoid as it did in the previous example. However, it still has a noticeable effect, as can be seen from the metric results, and from figure 2.18.

2.5.2 Effect of bracing

Figure 2.20 shows a 3-DOF planar manipulator. This example was first posed by Harry West [12] to illustrate how bracing could improve the load bearing ability of a simple planar manipulator. The idea was to have this manipulator pick up a load and move it horizontally.

For this example, the link lengths of the robot arm are all 1, and the

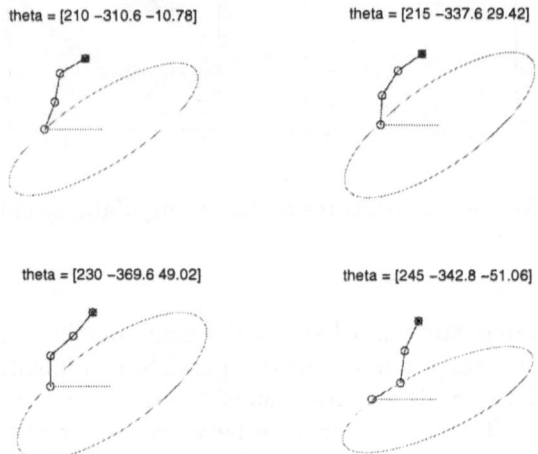

Figure 2.19: Four Different Postures of the Arm.

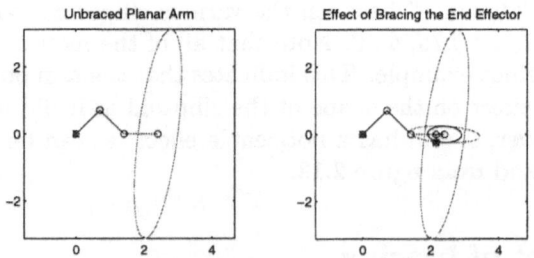

Figure 2.20: Effect of adding a brace on the load-bearing ability of a planar arm.

joint angles are $[45 \ -90 \ 45]^T$. The Jacobian for the unbraced arm is:

$$J_1 = \begin{bmatrix} 0 & 0.7071 & 0 \\ 2.4142 & 1.7071 & 1 \end{bmatrix} \qquad (2.37)$$

The large ellipsoid in the first part of the figure is the manipulability ellipsoid for the unbraced arm. The ellipsoid indicates that the arm configuration is good for motions, but poor for applying force (i.e. lifting objects) in the vertical direction.

To improve the performance of the arm, West proposed that a brace be mounted to the robot, near the end-effector. This brace would rest on the horizontal surface that the load rested on, and would support the arm. This brace could slide along the surface, and would also allow the robot arm to rotate about the point of contact between the brace and the horizontal surface.

The height of the brace was 0.25, and it was located 0.25 units from the end-effector. The motions which the brace allows make it equivalent to a two-link arm with a translational and a rotational joint, whose base is located in the same place as that of the brace itself [12]. Therefore, the Jacobian for the brace is:

$$J_2 = \begin{bmatrix} 1 & -0.25 \\ 0 & 0.25 \end{bmatrix} \qquad (2.38)$$

The smaller ellipsoid shown in the second part of figure 2.20 is the manipulability ellipsoid for the brace. The shape of the brace's ellipsoid indicates that the brace has greater force bearing capability in the vertical direction, but will readily allow motion in the horizontal direction.

Because the brace is attached to the robot arm, it can be treated as a rigid grasp (H^T does not exist). Let v_T be the linear end-effector velocity of the robot arm. The ellipsoid for the braced arm indicates that it has much better load-bearing capacity in the vertical direction than the unbraced arm, while it has retained nearly all of its ability to move in the horizontal direction. Thus, the overall effect of this brace is to drastically improve the lifting capability of the robot arm for this specific task.

It should be noted that the ellipsoid for the whole system is smaller than the ellipsoid for either arm taken individually. This makes sense; because of the kinematic constraint which each arm imposes upon the other, the arms restrict each other's motion. This effect can be seen in the reduced size of the ellipsoid.

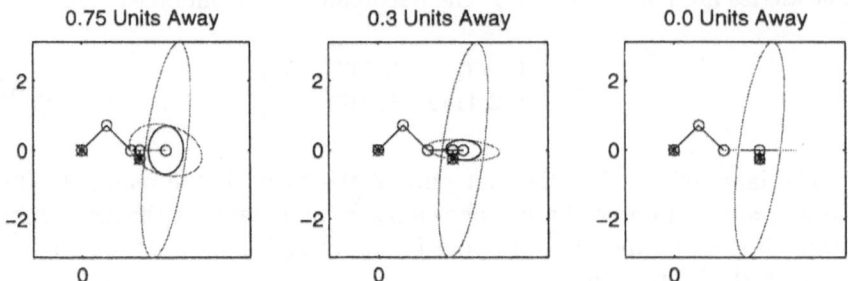

Figure 2.21: Effect of brace location on the manipulability ellipsoid.

2.5.3   Effect of brace location

Returning to the example shown in figure 2.20, it is reasonable to ask what gains can be achieved by altering the location brace on the robot arm.

¿From a load-bearing standpoint, the velocity ellipsoid of the braced arm system should be a horizontal line, (a degenerate ellipsoid) permitting only horizontal motion. However, because the brace has to be fixed somewhere, the brace will act as a fulcrum about which the last link of the arm can pivot. The weight of the load being lifted must be counteracted by the joints of the arm. Thus, the closer the brace is to the end-effector, the larger the load that the arm should be able to bear.

Figure 2.21 shows the effect of moving the brace closer to the end-effector of the robot. As the brace is placed closer to the end-effector, the ellipsoid of the braced system becomes shorter, indicating that the system is less able to move in the vertical direction, but more able to apply force in the vertical direction.

In the last part, the brace is exactly under the end-effector, and the system ellipsoid is degenerate, allowing only horizontal motion. In this situation, the load bearing ability of the braced arm would be (theoretically) infinite, since the load would be applying a force directly upon the kinematic structure of the bracing links, instead of on the joints of the main arm. However, there is a problem with placing the brace in this location. By having the brace directly underneath the end-effector, the robot end-effector no longer can change its height to *pick up* the workpiece. Thus, in addition to improving the manipulability of the system, the brace location must also allow for the task to be accomplished.

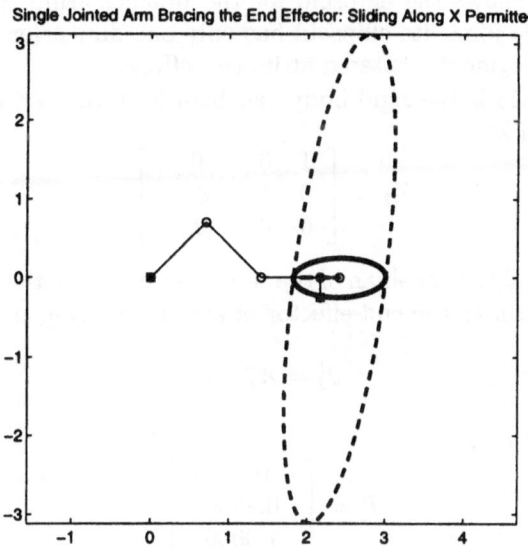

Figure 2.22: Effect of grasp contact type on the manipulability ellipsoid.

2.5.4 Effect of brace contact type

In [12], West modeled the braces he used as robot arms. In the example of the robot trying to lift a load (figure 2.20), the brace was modeled as a 2-jointed arm, with a prismatic and a revolute joint. However, the brace was in reality attached to the last link of the robot arm.

An alternative way of bracing a robot arm would be to have a single jointed, single link arm, upon which the first arm would rest its last link. This model more closely resembles the way that human arms are used to brace each another - each arm is separate, and the end-effectors (hands) are used to grasp and support objects. Figure 2.22 depicts this scenario.

As before, the Jacobian of the main arm is:

$$J_1 = \begin{bmatrix} 0 & 0.7071 & 0 \\ 2.4142 & 1.7071 & 1 \end{bmatrix} \tag{2.39}$$

As in West's example, the brace is 0.25 units tall, located 0.25 units behind the end-effector of the main arm. In this case, the bracing arm has only one (revolute) joint, so the Jacobian of the bracing arm is:

$$J_2 = \begin{bmatrix} -0.25 \\ 0 \end{bmatrix} \tag{2.40}$$

Figure 2.22 shows the ellipsoid for the bracing arm. Since the brace has only a single joint, its ellipsoid has only one dimension, and is thus a horizontal line segment, centered at its end-effector.

The matrix A_2 is the rigid body Jacobian from the end effector of arm 1 to that of arm 2:

$$A_2 = \begin{bmatrix} 1 & 0 & 0 \\ 0 & 1 & -0.25 \\ 0 & 0 & 1 \end{bmatrix} \tag{2.41}$$

We can extend the Jacobian of the second arm to map the joint velocities of the bracing arm to the end-effector of arm 1, by using the equation:

$$J_2' = A_2^{-1} J_2 \tag{2.42}$$

which yields the result:

$$J_2' = \begin{bmatrix} -0.2500 \\ 0.2500 \\ 1.0000 \end{bmatrix} \tag{2.43}$$

$$H_2^T = \begin{bmatrix} 1 \\ 0 \\ 0 \end{bmatrix} \tag{2.44}$$

It is also necessary to translate H_2^T to the point v_T (the end effector of the main arm), in order to maintain consistency in the equations. We can do this in the same manner as the Jacobian:

$$H_2'^T = A_2^{-1} H_2^T = \begin{bmatrix} 1 \\ 0 \\ 0 \end{bmatrix} \tag{2.45}$$

Since the main arm's grasp is rigid, H_1^T is nonexistent. H_2^T is a sliding contact in the x direction. The ellipsoid shown in Figure 2.22 with a solid line is the multiple-arm ellipsoid. Note that while the ellipsoid of the bracing arm is degenerate, the multiple-arm ellipsoid is not. A comparison of figures 2.22 and 2.20 shows that the multiple-arm ellipsoids for both systems are quite similar in size and shape.

Using the metrics presented earlier in this chapter, we find the "distance" between the ellipsoids to be: $\alpha = 0.1003$, $\beta = 0.0611$, $\gamma = 0.0913$, and $\delta = 0$. Thus, translationally, the ellipsoids are identical (as expected). Rotationally, scalewise, and shapewise, the differences are quite small. Such a result would be expected, since the bracing arms are similar in nature and location.

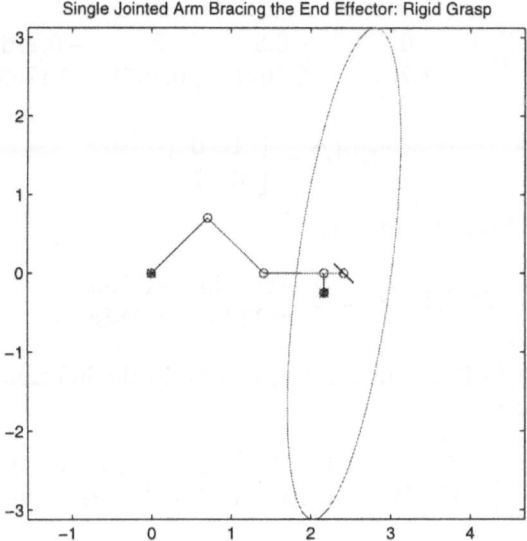

Figure 2.23: Effect of grasp contact type on the manipulability ellipsoid.

If the sliding contact is replaced by a rigid contact, the ellipsoid becomes degenerate (see Figure 2.23), indicating that motion is only permitted along a line.

As before, the system Jacobian is:

$$J = \begin{bmatrix} 0 & 0.7071 & 0 & 0 \\ 2.4142 & 1.7071 & 1 & 0 \\ 0 & 0 & 0 & -0.25 \\ 0 & 0 & 0 & 0.25 \end{bmatrix} \tag{2.46}$$

But in this case, since the grasp type of the bracing arm is rigid, H^T is nonexistent.

$$A = \begin{bmatrix} 1 & 0 \\ 0 & 1 \\ 1 & 0 \\ 0 & 1 \end{bmatrix} \tag{2.47}$$

Following the same calculation procedure, we obtain:

$$C_1 = (G^T)^+ J_h = \begin{bmatrix} 0 & 0.3536 & 0 & -0.125 \\ 1.2071 & 0.8536 & 0.5 & 0.125 \end{bmatrix} \tag{2.48}$$

$$C_2 = \widetilde{G}^T J_h = \begin{bmatrix} 0 & -0.5 & 0 & -0.1768 \\ -1.7071 & -1.2071 & -0.7071 & 0.1768 \end{bmatrix} \qquad (2.49)$$

$$\Omega^{-1/2} = \begin{bmatrix} 1 & 0 \\ 0 & 1 \end{bmatrix} \qquad (2.50)$$

And finally, we obtain the result:

$$C_1 \widetilde{C}_2 \Omega^{-1/2} = \begin{bmatrix} -0.1305 & -0.1836 \\ 0.1305 & 0.1836 \end{bmatrix} \qquad (2.51)$$

Applying the SVD to this matrix, we obtain the information about the multi-arm ellipsoid:

$$U = \begin{bmatrix} -0.7071 & -0.7071 \\ 0.7071 & -0.7071 \end{bmatrix} \quad \Sigma = \begin{bmatrix} 0.3186 & 0 \\ 0 & 0 \end{bmatrix} \qquad (2.52)$$

Comparing this figure with 2.22 shows the drastic effect that the grasp type may have on the system manipulability. ($\alpha = 1.1170$, $\beta = 0.4196$, $\gamma = 0.2744$, $\delta = 0$.) Note that the metric results indicate a much greater difference than was noted between West's example and the sliding contact result.

2.6 Comparison of manipulability ellipsoids

In order to use ellipsoids to guide the selection of robot pose, grappling point, and contact type, it is necessary to measure the "distance" of a given ellipsoid to a desired ellipsoid. In this section, we consider several possible metrics for ellipsoids. In addition, we also consider the special case of degenerate ellipsoids.

Metrics involving ellipsoids have not received much attention in the literature. Several groups [13, 14, 15] have been concerned with using ellipsoids as an aid in robot kinematic design. In [16], the manipulability ellipsoid is used to specify the desired manipulability of the robot arm. Their approach was to make the desired ellipsoid scalable, and they sought the largest desired ellipsoid which would fit inside the actual ellipsoid of the arm. A maximum value was achieved when the desired ellipsoid was the same size and shape as the actual ellipsoid. In [11, 17], the manipulability ellipsoid is also used to specify the desired performance of the robot arm, and the desired ellipsoid is compared with the actual ellipsoid of the robot. He proposed two different methods of comparing ellipsoids [11]: the volume

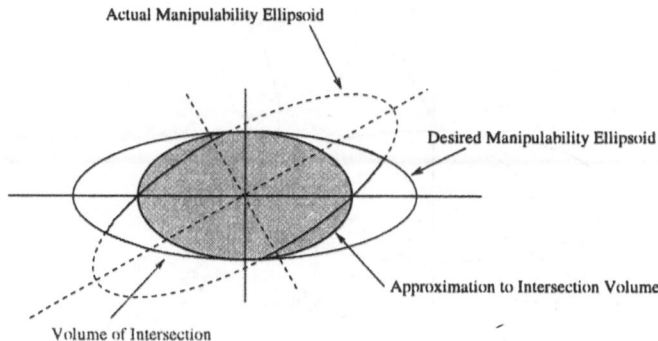

Figure 2.24: Volume of intersection between two ellipsoids.

of intersection and a "shape discrepancy" measure, along the principal axes of the ellipsoid. Neither of these measures is a true metric, however.

The first measure to compare two ellipsoids which Lee proposed was their volume of intersection. Figure 2.24 shows a typical example in two dimensions. One benefit to such a method is that it is readily understandable. However, the intersection of two ellipsoids does not usually result in an ellipsoid, but in a more complicated shape which is difficult to describe mathematically.

Because of this complexity, Lee approximated the volume of intersection by a new ellipsoid, whose principal axes were determined from the principal axes of the desired ellipsoid, or from the intersection of the principal axes of the desired ellipsoid with the boundary of the actual ellipsoid, whichever was shorter.

The volume of an m-dimensional ellipsoid is straightforward to compute [18]:

$$\text{vol} = d\,\sigma_1\,\sigma_2\,\sigma_3\,\ldots\,\sigma_m \tag{2.53}$$

where $\sigma_1,\ldots,\sigma_m$ are the singular values of the Jacobian, and d is a constant given by

$$d = \begin{cases} (2\pi)^{m/2}/(2\cdot4\cdot6\cdot\ldots(m-2)\cdot m) & m \text{ even} \\ 2\,(2\pi)^{(m-1)/2}/(1\cdot3\cdot5\ldots(m-2)\cdot m) & m \text{ odd} \end{cases} \tag{2.54}$$

For ease of computation, it may not be necessary to calculate d. The product of the singular values of the arm Jacobian will yield a result which is proportional to the true volume of the ellipsoid.

There are a several drawbacks to using the approximation method. First, it uses an estimate of the volume, rather than the volume itself.

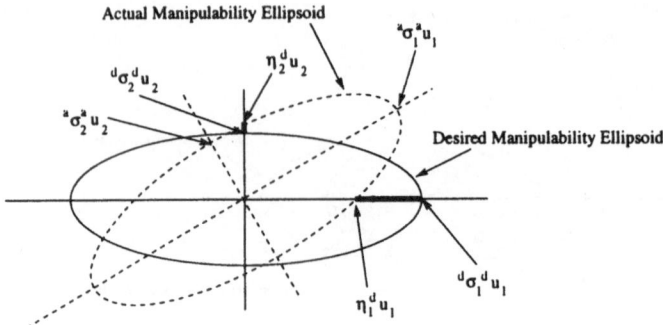

Figure 2.25: Shape discrepancy between two ellipsoids.

There is no indication that the approximation will be close to the true volume of intersection. Second, it is not clear why the approximation ellipsoid is defined the way it is; an equally valid (but different) ellipsoid could be obtained by using the principal axes of the actual ellipsoid, instead of those of the desired ellipsoid.

Still another drawback is that this method is not a metric, and does not result in a "distance" measurement. The measure results in a maximum when the desired ellipsoid is completely contained in the actual ellipsoid. It is possible that many differently-shaped manipulability ellipsoids could contain a given desired ellipsoid; the volume of intersection method would return the same result for each of these ellipsoids. It would not be possible to identify any particular manipulability ellipsoid as being superior to the others.

Finally, this method cannot handle degenerate ellipsoids. It is easier for people to specify tasks as simple ellipsoids; one of the simplest is a line - which can be viewed as an ellipsoid with only one non-zero principal axis. (Such a desired ellipsoid would indicate that motion was only desired along a single direction.) The volume method would always return zero in such a case, providing no useful information.

Shape discrepancy

The second method proposed by Lee compared the distance from the center of the ellipsoids to their edges, along the principal axes of the desired ellipsoid. See figure 2.25.

The darkened lines indicate the distances which were calculated as part of the shape discrepancy measure. The complete measure which Lee used was the reciprocal of the sum of squares of these lengths:

$$\text{discrepancy} = 1/\sum_{i=1}^{m}(^{d}\sigma_i - \eta_i)^2 \qquad (2.55)$$

This measure is more useful than the volume of intersection, and provides information on the shape difference of the ellipsoids. It does not fail with degenerate ellipsoids.

However, this measure still has flaws. First, the measure as given will tend to infinity as the actual ellipsoid tends to the desired ellipsoid. This can be remedied by not taking the reciprocal of the sum of squares; however, even this modified measure still is not a metric, as the result from the actual ellipsoid to the desired one is different than if the roles of the ellipsoid were reversed.

Intuitively, there are several ways to distinguish between two ellipsoids:

- translation: the centers of the ellipsoids are not located at the same point in space.

- rotation: the corresponding principal axes of the ellipsoids point in different directions.

- scaling: the corresponding principal axes of the ellipsoids differ by a constant scale factor; the principal axes of each ellipsoid have multiplicative relationships between the axes, and these relationships hold for both ellipsoids, but the lengths of the corresponding axes are different.

- shape discrepancy: the relationship between the lengths of the principal axes in each ellipsoid differs, resulting in a different shape for each ellipsoid.

Any metric function we choose must be able to handle the above cases. In addition, the metric function must be able to handle degenerate ellipsoids - which occurs when one or more principal axes of an ellipsoid has zero length. One way of constructing such a metric is to determine a metric for each of the above cases individually.

Define the following attributes of an ellipsoid:

- position of the center of the ellipsoid c.

- m unit vectors pointing along the ellipsoid's principal axes $u_1, \ldots, u_m$.

- m lengths of the principal axes $\sigma_1, \ldots, \sigma_m$.

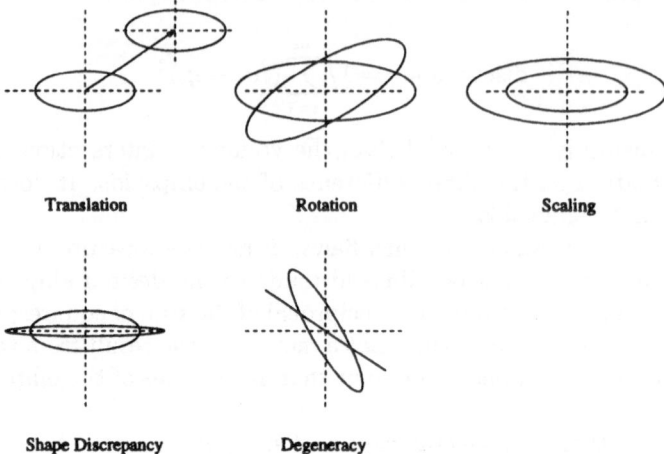

Figure 2.26: Comparison between ellipsoids

The following functions are proposed for comparing ellipsoids. Proofs showing that they are indeed metrics can be found in [19]. (Strictly speaking, the functions are semi-metrics; however, each function can be made into a metric by considering all ellipsoids which result in the function being zero as belonging to the same equivalent class.)

translation:

$$\delta(\mathcal{E}_1, \mathcal{E}_2) = \|c_2 - c_1\| = \sqrt{(c_2 - c_1)^T (c_2 - c_1)} \qquad (2.56)$$

where c_i is the location of the center of ellipsoid i.

rotation:

$$\alpha(\mathcal{E}_1, \mathcal{E}_2) = \sum_{i=1}^{m} \sqrt{1 - \cos \phi([u_1^{(i)}], [u_2^{(i)}])} \qquad (2.57)$$

where $[u_1^{(i)}]$ and $[u_2^{(i)}]$ are the ith (nondirectional) principal axes of ellipsoids 1 and 2, respectively. That is, $[u^{(i)}]$ is the union of the vectors $u^{(i)}$ and $-u^{(i)}$. The function $\phi([u_1], [u_2]) = \min\{\theta(u_1, u_2), \theta(-u_1, u_2)\}$.

scaling:

$$\gamma(\mathcal{E}_1, \mathcal{E}_2) = |\sigma_1^{(1)} - \sigma_1^{(2)}| \qquad (2.58)$$

shape discrepancy:

$$\beta(\mathcal{E}_1, \mathcal{E}_2) = \sum_{i=1}^{m} |\varepsilon_i^{(1)} - \varepsilon_i^{(2)}| \qquad (2.59)$$

where:

$$\varepsilon_i^{(j)} = \begin{cases} \frac{\sigma_i^{(j)}}{\sigma_1^{(j)}} & \text{if } \sigma_1^{(j)} > 0 \\ 0 & \text{if } \sigma_1^{(j)} = 0 \end{cases} \qquad (2.60)$$

A weighted sum of these functions is itself a metric function, provided that the weights are nonnegative. (If the weights are all positive, the weighted sum is a true metric.) This formulation has the advantage that certain aspects of the "distance" between ellipsoids can be emphasized if it would help in optimizing a certain task. For example, if the orientation of the ellipsoid was found to be more important to some task than the scaling of the ellipsoid, the orientation component could have a larger weighting in the calculation of the composite metric.

It is important to note that all of the component functions still return useful results, even if the lengths of one or more of the principal axes of the ellipsoids is zero (where the ellipsoid is degenerate). This is because none of the functions depend upon the lengths of the principal axes being non-zero.

To provide some general insight into the values returned from these metrics, Figures 2.27–2.29 illustrate the results of the rotational, scale and shape discrepancy metrics. The translational metric is not shown, as it is fairly well understood. Note that while the metric functions work for ellipsoids of other dimensions, two-dimensional ellipsoids are used in these examples because they are the easiest to visualize.

Figure 2.27 shows a two-dimensional reference ellipsoid (depicted by a solid line) and several other ellipsoids (shown with dotted lines) rotated in increments of 30°. The corresponding values of α are shown next to these ellipsoids. Note that α achieves a maximum value of 2 when the ellipsoids are 90° apart. A check of the formula for α shows that in general, the maximum value it can attain is m - which would happen when all m corresponding principal axes of the ellipsoids being compared are at right angles to each other.

Figure 2.28 shows a reference ellipsoid, along with several ellipsoids which have different shapes. The values of β are shown next to each ellipsoid. Note that the β increases as the shape of the ellipsoid differs more greatly from the initial form. The maximum value that β might attain is m - but this will only happen under special circumstances. For this to occur, the one ellipsoid would have to be a ball (all principal axes are of the same length), while the other ellipsoid would have to be a point (all principal axes are of zero length). More generally, the maximum attainable value of β would be $\sum_1^m \sigma_i^{(1)}/\sigma_1^{(1)}$, which would occur when the second ellipsoid was a point.

Figure 2.29 shows some results of the scaling metric. As the ellipsoids get larger or smaller than the reference ellipsoid (where size is determined by the largest singular value), γ becomes larger. As the second ellipsoid increases in size, $\gamma \to \infty$.

The range of values which the metrics can attain is summarized below:

$$
\begin{aligned}
\alpha &: \quad 0 \longleftrightarrow m \\
\beta &: \quad 0 \longleftrightarrow m \\
\gamma &: \quad 0 \longleftrightarrow \infty \\
\delta &: \quad 0 \longleftrightarrow \infty
\end{aligned}
\tag{2.61}
$$

Figures 2.30 – 2.32 show some two-dimensional ellipsoids which differ in more than one aspect, and the resultant metric values. Figure 2.30 shows two identically shaped ellipsoids which are rotated and translated relative to each other. Note that the appropriate metrics indicate this difference; however, the shape discrepancy and scale metrics return zero (because the ellipsoids have the same shape and scale).

Figure 2.31 show two ellipsoids of different shape, centered at the same point. They are rotated slightly with respect to each other. The ellipsoids' longest principal axes are identical in length. In this case, the translation metric and the scale metric return zero, while the rotational metric and the shape discrepancy metric return non-zero values, indicating a difference between the two ellipsoids.

Figure 2.32 shows two ellipsoids which have the same translational and rotational position, but which are shaped and scaled differently. Again, only the metrics which pertain to shape and scale return non-zero values.

It may be the case that which metrics are emphasized will be dependent upon the task being performed. For example, in controlling a pool cue, the exact rotation of the ellipsoid may not be important, as long as the robot can readily move the cue in the needed direction. Translation would be important, since a translational error would indicate that the tip of the pool cue is positioned incorrectly. The shape of the ellipsoid would also be important, because we would want to most readily move in the desired direction, while resisting disturbances in other directions.

If the robot has a different task, such as picking up and moving an object, other metrics may be more important. If the exact location on the

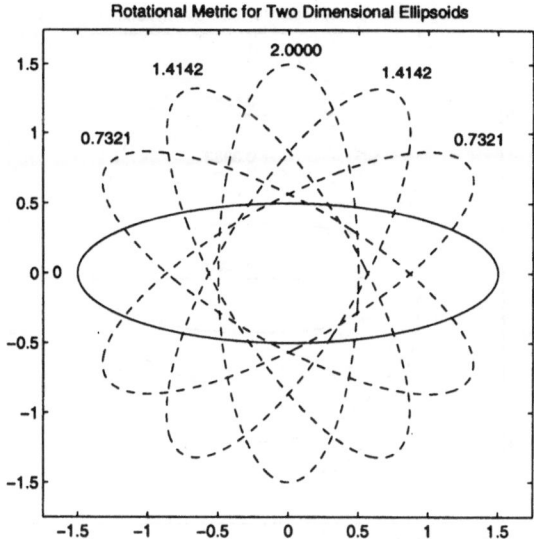

Figure 2.27: Metric example 1: The values of α for several different orientations.

object that the gripper contacts is not critical, then the translational metric may not be that important. The shape of the ellipsoid would be important (and rotation somewhat less so), since the end-effector should have good mobility in the direction the object should be moved, but should have very short dimension in the vertical axis (the direction force must be applied in order to lift the object).

2.7 Conclusions

This paper generalizes the velocity and force manipulability to general constrained multibody systems. Such systems include simple closed kinematic chain as two arms jointly holding a payload, multiple kinematic chains as in multi-finger grasping, and more complex structures as multiple Stewart Platforms. We have extended the concept of stable grasp and manipulable grasp in the multi-finger grasp literature to general mechanisms and provide necessary and sufficient conditions for their verifications. In general, unstable (or nearly unstable) configurations need to carefully considered in the kinematic analysis, otherwise there may be uncontrolled motion or large joint loading. We have also shown that that multiple arm manipulability

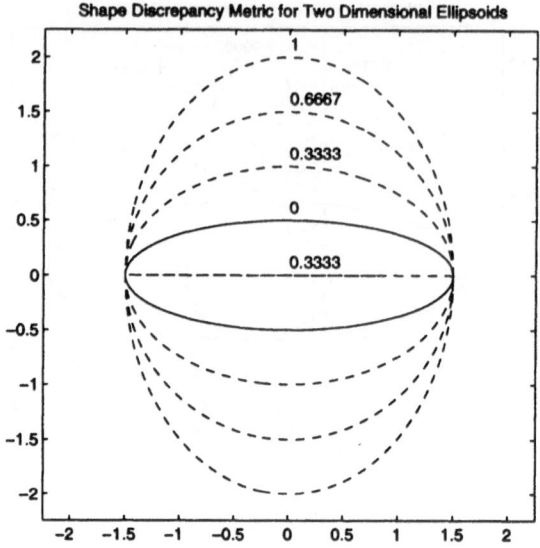

Figure 2.28: Metric example 2: The effect of altering the shape of the ellipsoid on β.

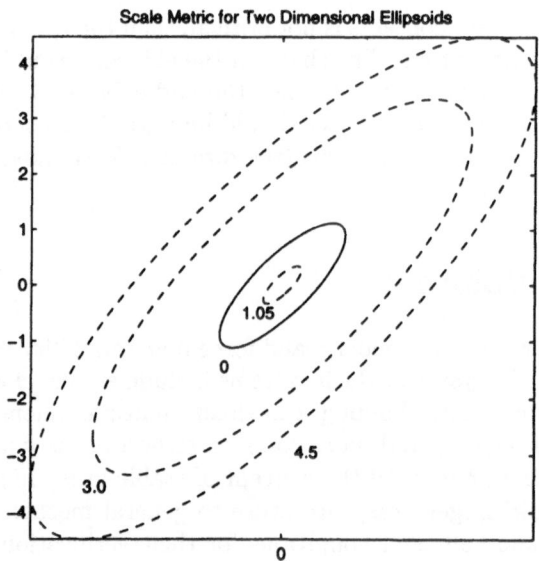

Figure 2.29: Metric example 3: The values of γ for a series of scaled ellipsoids.

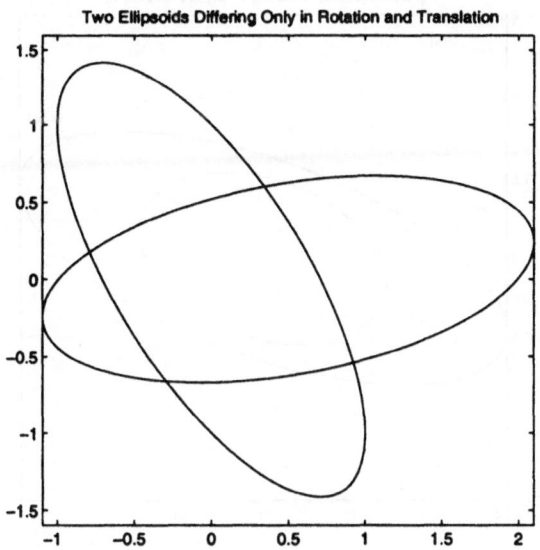

Figure 2.30: Metric example 4: $\alpha = 1.5874$, $\beta = 0$, $\gamma = 0$, $\delta = 0.5$.

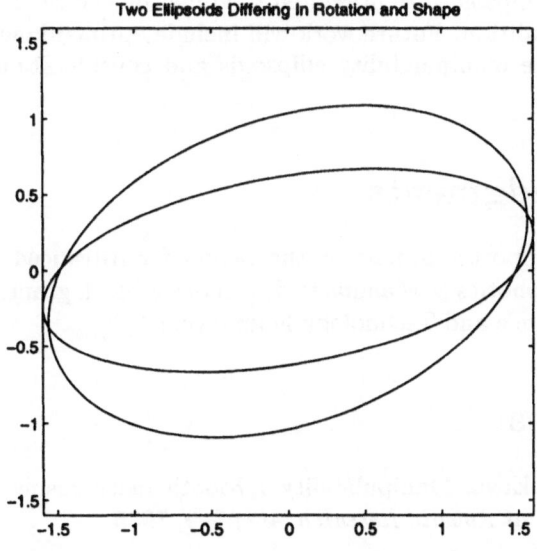

Figure 2.31: Metric example 5: $\alpha = 0.2465$, $\beta = 0.361$, $\gamma = 0$, $\delta = 0$.

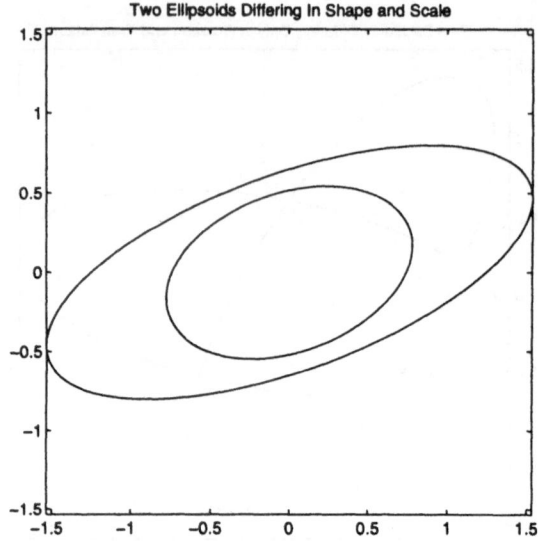

Figure 2.32: Metric example 6: $\alpha = 0$, $\beta = 0.2361$, $\gamma = 0.809$, $\delta = 0$.

can be significantly modified through the bracing by arms. Several metrics for comparing ellipsoids are presented to guide the choice of brace location and contact condition. Future work will include optimal kinematic synthesis based on the manipulability ellipsoids and consideration of dynamics and control.

Acknowledgments

This work is supported in part by the Center for Advanced Technology in Automation, Robotics & Manufacturing under a block grant from the New York State Science and Technology Foundation.

References

[1] T. Yoshikawa. Manipulability of robotic mechanisms. *International Journal of Robotic Research*, 4(2):3–9, 1985.

[2] F. Park. Manipulability of closed kinematic chains. submitted to ASME Journal on Mechanical Design, 1997.

[3] P. Chiacchio, S. Chiaverini, L. Sciavicco, and B. Siciliano. Reply to "comments on 'global task space manipulability ellipsoids for multiple-arm systems' and further considerations". *IEEE Transactions on Robotics and Automation*, 9(2):235–236, April 1993.

[4] A. Bicchi, C. Melchiorri, and D. Balluchi. On the mobility and manipulability of general multiple limb robots. *IEEE Transactions on Robotics and Automation*, 11(2):215–228, April 1995.

[5] M. Uchiyama and P. Dauchez. Symmetric kinematic formulation and non-master/slave coordinated control of two-arm robots. *Advanced Robotics: The international Journal of the Robotics Society of Japan*, 7(4):361–383, 1993.

[6] V.R. Kumar and K.J. Waldron. Force distribution in closed kinematic chains. *IEEE Journal of Robotics and Automation*, 4(6):657–664, December 1988.

[7] C.A. Klein and S. Kittivatcharapong. Optimal force distribution for the legs of a walking machine with friction cone constraints. *IEEE Transaction on Robotics and Automation*, 6(1):73–85, February 1990.

[8] P. Chiacchio, S. Chiaverini, L. Sciavicco, and B. Siciliano. Global task space manipulability ellipsoids for multiple-arm systems. *IEEE Transactions on Robotics and Automation*, 7(5):678–685, October 1991.

[9] C. Melchiorri. Comments on 'global task space manipulability ellipsoids for multiple-arm systems' and further considerations. *IEEE Transactions on Robotics and Automation*, 9(2):232–235, April 1993.

[10] A. Bicchi and C. Melchiorri. Manipulability measures of cooperating arms. In *Proceedings of the 1993 American Controls Conference*, pages 321–325, San Francisco, CA, June 1993.

[11] S. Lee. Dual redundant arm configuration optimization with task-oriented dual arm manipulability. *IEEE Transactions on Robotics and Automation*, 5(1):78–97, February 1989.

[12] H. West. *Kinematic Analysis for the Design and Control of Braced Manipulators*. PhD thesis, Massachusetts Institute of Technology, 1984.

[13] C. Gosselin. The optimum design of robotic manipulators using dexterity indices. In *Robotics and Autonomous Systems 9*, pages 213–226. Elsevier Science Publishers, 1992.

[14] F. Ranjbaran, J. Angeles, and A. Kecskemethy. On the kinematic conditioning of robotic manipulators. In *Proceedings of the 1996 IEEE International Conference on Robotics and Automation*, pages 3167–3172, Minneapolis, MN, 1996.

[15] C. A. Klein and B. E. Blaho. Dexterity measures for the design and control of kinematically redundant manipulators. *International Journal of Robotic Research*, 6(2):72–83, summer 1987.

[16] Z. Li, P. Hsu, and S. Sastry. Grasping and coordinated manipulation by a multifingered robot hand. *International Journal of Robotic Research*, 8(4):33–50, August 1989.

[17] S. Lee and S. Kim. A self–reconfigurable dual–arm system. In *Proceedings of the 1991 IEEE International Conference on Robotics and Automation*, pages 164–169, Sacramento, CA, April 1991.

[18] T. Yoshikawa. Analysis and control of robot manipulators with redundancy. In M. Brady and R. Paul, editors, *Robotics Research: The First International Symposium*, pages 735–747. MIT Press, 1984.

[19] L. Wilfinger. Robotic bracing. Doctoral Candidacy Proposal, 1997.

Chapter 3

Kinematic control of dual-arm systems

This chapter focuses on the control problem for cooperative manipulator systems in the framework of kinematic control. The control problem is solved in two stages: first, an inverse kinematics problem is solved to transform cooperative task variables into the corresponding joint variables for all the manipulators constituting the cooperative system; then, the obtained joint variables are fed to a suitable joint space control scheme. A useful feature of the chosen approach is that coordination is solved at inverse kinematics level while arm interactions can be handled at joint control level.

An effective formulation is presented which fully characterizes a coordinated motion task in terms of a set of meaningful position and orientation variables. A closed-loop algorithmic approach for the inverse kinematics problem is pursued based on differential kinematics mappings; this also allows handling of kinematic redundancy and singular configurations.

A joint-space control scheme based on kineto-static filtering of the joint errors is presented which is aimed at canceling out the internal force at steady state without using force measurements. Nevertheless, when force sensors are available, feedback of the internal forces can be added to improve performance of the system. Stability analysis is provided and asymptotic stability of a set of equilibrium points is demonstrated. The case of imperfect compensation of the gravity terms is also discussed.

3.1 Introduction

The two main goals of the control problem for cooperative manipulators are coordinated motion of multiple-arm systems and handling of internal forces arising from arm interactions through a commonly held object. In order to achieve these goals, a number of major aspects must be taken into account.

First, specification of the cooperative task requires definition of a set of variables which are meaningful to the user and still allow a complete description of the objectives of the cooperation.

Then, the control scheme should be robust to the occurrence of singular configurations of the cooperative system; moreover, the capability of exploiting redundant degrees of freedom possibly present in the system should be provided.

When the cooperative manipulator system results from grouping individual robot systems, i.e. arm plus controller, it would be desirable to adopt a cooperative control scheme that takes advantage at the most of the decentralized structure inherent to the available hardware.

Looking forward to industrial application of cooperative manipulator systems, the use of force sensors should not be crucial to functioning of the control scheme. In this respect, it is assumed that reduced performance is acceptable for a control scheme not requiring force measurement.

In view of the above aspects, we have chosen to deal with the control problem for cooperative manipulator systems by resorting to a *kinematic control* approach. This means that the control problem is solved in two stages: first, an inverse kinematics problem is solved to transform cooperative task variables into the corresponding joint variables for all the manipulators constituting the cooperative system; then, the obtained joint variables are fed to a suitable joint space control scheme. A useful feature of the chosen approach is that coordination is solved at inverse kinematics level while arm interactions can be handled at joint control level.

To set up an inverse kinematics for cooperative manipulator systems it is first necessary to find a task description that allows specification of coordinated motion. To this purpose, it should be obvious that taking the end-effector position and orientation of each manipulator as task variables is inadequate, since the system would be regarded as composed by independent manipulators and coordination management would be left to the user. An effective formulation is presented in this chapter which fully characterizes a coordinated motion task in terms of a set of meaningful position and orientation variables.

Finding closed-form solutions to the inverse kinematics problem is possible only for special manipulator geometries and simple coordination tasks.

An algorithmic approach shall be pursued instead based on differential kinematics mappings; this also allows handling of kinematic redundancy and singular configurations. We resort to a closed-loop inverse kinematics algorithm which does not suffer from numerical drift typical of open-loop algorithms.

If an independent joint control law is used as second stage of the kinematic control scheme, arising of internal forces is expected. These forces originate from different sources; namely, joint tracking errors causing violation of closed-chain constraints, joint trajectories not consistent with the geometry of the grasp, non-compensated dynamics. The effect of the last source can be effectively reduced only by resorting to model-based control schemes; however, these schemes are computationally demanding for cooperative systems where also the (possibly unknown) dynamics of the object plays a significant role.

To retain the simplicity of a scheme without dynamic compensation we present a control scheme based on kineto-static filtering of the joint errors aimed at reducing the building of internal forces. A simple PD-type control law is described which is shown to cancel out the internal force at steady state. Remarkably, this is obtained without using force measurements. Nevertheless, when force sensors are available, feedback of the internal forces can be added to improve performance of the system and to achieve internal force regulation.

Stability analysis is developed for the equilibrium points of the cooperative system under the proposed joint-space control laws. By applying the global invariant set theorem in a singularity-free region, asymptotic convergence to the invariant set constituted by the equilibrium points is demonstrated and local stability of the minimum norm equilibrium points is analyzed. The analysis is extended also to the case of imperfect compensation of the gravity terms.

3.2 Cooperative task description

Most available task formulations are based on a global description of the system through the use of the so-called grasp matrix assuming that an object is commonly held by the manipulators [1,2,3,4]; the result is a complete description of the task at differential kinematics level. Position task variables can be easily found by integration of linear velocities while a problem arises for orientation task variables, in view of the non-integrability of angular velocities. It might be argued that a description of orientation could be found by means of a minimal representation, e.g. Euler angles, but the resulting variables would not allow a clear specification of a coordination task for the user. To overcome this problem, a description of the coordinated motion task will be sought in terms of a set of meaningful position and orientation variables [5].

In the present work, a system of two cooperative manipulators, namely manipulator 1 and manipulator 2, will be considered. In the remainder, the subscript i will denote quantities referred to manipulator i. Let Σ_i denote the frame attached to the end effector of manipulator i; quantities referred to Σ_i will be denoted by the superscript i. Moreover, all quantities referred to a common base frame Σ_b will be denoted by the superscript b.

Then, let p_i^b be the (3×1) vector denoting the end-effector position as the origin of Σ_i. Let also R_i^b be the (3×3) rotation matrix expressing the end-effector orientation, i.e., its columns represent the unit vectors of Σ_i.

In order to establish the sought task description, a suitable frame is to be introduced to specify coordinated motion of the two-manipulator system. Let such frame be termed as absolute frame and denoted by Σ_a; quantities referred to the absolute frame will be denoted by the superscript a.

The origin of the absolute frame is called *absolute position* of the cooperative system and can be expressed as a function of the positions of the two end effectors. One simple choice is

$$p_a^b = \frac{1}{2}(p_1^b + p_2^b). \tag{3.1}$$

In order to define the absolute orientation of the system, consider the matrix operator $R_k(\vartheta)$ expressing the rotation by the angle ϑ about the axis aligned with the unit vector $k = [\, k_x \quad k_y \quad k_z \,]^{\mathrm{T}}$ [6]. Then, the rotation matrix giving the *absolute orientation* can be defined as

$$R_a^b = R_1^b R_{k_{12}^1}^1(\vartheta_{12}/2), \tag{3.2}$$

where k_{12}^1 and ϑ_{12} are respectively the unit vector and the angle that realize the rotation described by R_2^1, i.e., the orientation of frame 2 with respect to frame 1. Therefore, the above choice corresponds to make a rotation about axis k_{12}^1 by an angle which is half the angle needed to align R_2^b with R_1^b.

The absolute position and orientation describe the task in terms of the composition of the position and orientation of the single manipulators and thus it is clear that there exist infinite end-effector configurations giving the same absolute position and orientation. Therefore, in order to fully describe a coordinated motion, the position and orientation of one manipulator relative to the other is also of concern.

The *relative position* between the two end effectors can be defined as

$$p_r^b = p_2^b - p_1^b. \tag{3.3}$$

The *relative orientation* between the two end effectors can be defined with reference to the end-effector frame of either manipulator —say the first one— in terms of the rotation matrix

$$R_r^1 = R_2^1. \tag{3.4}$$

To simplify specification of some coordination task, it might be appropriate to choose p_i^b and R_i^b other than the actual end-effector position and orientation of the two manipulators. This results in embedding proper constant transformation matrices in the direct kinematics of the two manipulators.

To be independent of the absolute motion of the system, it might be more convenient to specify the relative position with reference to the absolute frame, i.e., p_r^a. The relationship between p_r^a and p_r^b is given by

$$p_r^b = R_a^b p_r^a \tag{3.5}$$

with R_a^b as in (3.2).

A feature of the proposed formulation is that coordinated motion of the system is achieved without necessarily assuming that the two manipulators are kinematically constrained through the presence of an object between the two end effectors. Nevertheless, if the two end effectors hold a common object, general manipulation tasks can be described by the above formulation. For instance, if the task is to move a tightly grasped object without deforming it, a trajectory has to be assigned to p_a^b and R_a^b while p_r^a and R_r^1 have to be kept constant. Yet, if the task is to stretch, bend or shear the object, suitable trajectories have to be specified for the relative variables too.

3.3 Differential kinematics

Having established a task formulation for the direct kinematics of the two-manipulator system, it is useful to derive also the differential kinematics relating the coordinated (absolute and relative) velocities to the end-effector velocities of the two manipulators.

For each manipulator, the end-effector linear velocity is directly given as the time derivative of the position vector, that is $\dot{p}_i^b$. The end-effector angular velocity is given by the (3×1) vector ω_i^b, which is related to the time derivative of the rotation matrix R_i^b through the relationship

$$\dot{R}_i^b = S(\omega_i^b) R_i^b, \tag{3.6}$$

where $S(\cdot)$ is the (3×3) skew-symmetric operator performing the cross product.

The absolute linear velocity of the system is obtained as the time derivative of (3.1), i.e.,

$$\dot{p}_a^b = \frac{1}{2}(\dot{p}_1^b + \dot{p}_2^b). \tag{3.7}$$

Differentiating (3.2) with respect to time, using (3.6) and the relationship [6]

$$R_1^b S(\omega_{12}^1)(R_1^b)^{\mathrm{T}} = S(R_1^b \omega_{12}^1), \tag{3.8}$$

yields

$$S(\omega_a^b) R_a^b = S(\omega_1^b) R_a^b + \frac{1}{2} S(\omega_{12}^b) R_a^b, \tag{3.9}$$

where ω_{12}^b denotes the angular velocity of frame 2 with respect to frame 1.

From (3.9) it can be recognized that the absolute angular velocity is given by

$$\omega_a^b = \frac{1}{2}(\omega_1^b + \omega_2^b), \tag{3.10}$$

since $\omega_{12}^b = \omega_2^b - \omega_1^b$.

The relative linear velocity of the system is obtained as the time derivative of (3.3), i.e.,

$$\dot{p}_r^b = \dot{p}_2^b - \dot{p}_1^b. \tag{3.11}$$

If the relative position is expressed as p_r^a, the time derivative of (3.5) gives

$$\dot{p}_r^b = R_a^b \dot{p}_r^a + S(\omega_a^b) p_r^b \tag{3.12}$$

with ω_a^b as in (3.10).

Finally, differentiating (3.4) with respect to time and using (3.6) yields

$$S(\omega_r^1) R_r^1 = S(\omega_{12}^1) R_2^1, \tag{3.13}$$

and thus the relative angular velocity is

$$\omega_r^b = \omega_2^b - \omega_1^b, \tag{3.14}$$

which has been expressed in the base frame.

Algorithmic solutions to the inverse kinematics problem are based on the computation of the Jacobian associated with the task variables of interest. Since these variables have been expressed as a function of the position and orientation of the two end effectors, the sought Jacobian can be related to the Jacobians of the single manipulators.

Let n_i denote the number of joints of manipulator i and q_i be the $(n_i \times 1)$ vector of its joint variables. The geometric Jacobian $J_i^b(q_i)$ is the $(6 \times n_i)$

matrix relating the joint velocity vectors $\dot{q}_i$ to the linear and angular end-effector velocities in the base frame as

$$\begin{bmatrix} \dot{p}_i^b \\ \omega_i^b \end{bmatrix} = J_i^b(q_i)\dot{q}_i \qquad i = 1,2. \tag{3.15}$$

At this point, combining (3.7),(3.10) and taking into account (3.15) yields

$$\begin{bmatrix} \dot{p}_a^b \\ \omega_a^b \end{bmatrix} = J_a^b(q_1, q_2) \begin{bmatrix} \dot{q}_1 \\ \dot{q}_2 \end{bmatrix}, \tag{3.16}$$

where the $(6 \times (n_1 + n_2))$ *absolute Jacobian* matrix is defined as

$$J_a^b = [\tfrac{1}{2}J_1^b \quad \tfrac{1}{2}J_2^b]. \tag{3.17}$$

Further, combining (3.11),(3.14) and taking into account (3.15) yields

$$\begin{bmatrix} \dot{p}_r^b \\ \omega_r^b \end{bmatrix} = J_r^b(q_1, q_2) \begin{bmatrix} \dot{q}_1 \\ \dot{q}_2 \end{bmatrix}, \tag{3.18}$$

where the $(6 \times (n_1 + n_2))$ *relative Jacobian* matrix is defined as

$$J_r^b = [-J_1^b \quad J_2^b]. \tag{3.19}$$

3.4 Inverse kinematics algorithm

The inverse kinematics problem for a two-manipulator system can be stated as that to compute the joint variable trajectories corresponding to given co-ordinated motion trajectories for the absolute and relative task variables. Finding closed-form solutions is possible only for special manipulator geometries and simple coordination tasks, and thus an algorithmic approach shall be pursued.

An effective *inverse kinematics algorithm* is given by the closed-loop scheme based on the computation of the pseudoinverse of the manipulator Jacobian [7,8]. The joint velocity solution can be written in the general form

$$\dot{\bar{q}} = \bar{J}^\dagger(\bar{v}_d + \bar{K}\bar{e}) + (I - \bar{J}^\dagger\bar{J})\dot{\bar{q}}_0, \tag{3.20}$$

where $\bar{q}$ is a vector of joint variables, $\bar{J}$ is the Jacobian associated with the velocity mapping, $\bar{v}_d$ is the desired task velocity, $\bar{K}$ is a suitable diagonal positive gain matrix, $\bar{e}$ is the algorithmic error between the desired and current task variables, and $\dot{\bar{q}}_0$ is a vector of joint velocities that can be freely chosen.

Notice that the physical robot system is not involved since the algorithm only serves the purpose to invert the kinematics of the system along a given task trajectory, i.e., $\dot{\bar{q}}$ is given by (3.20), and $\bar{q}$ is computed by integrating the obtained $\dot{\bar{q}}$.

The closed-loop inverse kinematics algorithm based on (3.20) avoids the typical numerical drift of open-loop resolved-rate schemes [9]. The solution can be made robust with respect to singularities of J by resorting to a damped least-squares inverse of the matrix [10,11].

When $\bar{J}$ is square and full rank the pseudoinverse in (3.20) reduces to the inverse and the second term on the r.h.s. vanishes. If $\bar{J}$ has more columns than rows the system is *kinematically redundant*; in this case, the joint velocity vector $\dot{\bar{q}}_0$ can be suitably chosen to meet additional constraints besides the primary task. This is made possible since the term $(I - \bar{J}^\dagger \bar{J})$ is a projector onto the null space of $\bar{J}$ and thus the second term on the r.h.s. of (3.20) allows reconfiguration of the system without affecting the primary task [12].

It should be pointed out that kinematic redundancy of two-manipulator systems may be due either to the effective presence of additional joint variables —i.e., more than 6 degrees of freedom for either manipulator— or to relaxation of some coordination task variables. In [5] it has been show that non-tight grasps can be treated as a special case of kinematic redundancy.

The above algorithm can be applied to solve the inverse kinematics for the two-manipulator system at issue. In detail, define

$$\bar{q} = \begin{bmatrix} q_1 \\ q_2 \end{bmatrix}. \tag{3.21}$$

The task Jacobian is

$$\bar{J} = \begin{bmatrix} J_a^b \\ J_r^b \end{bmatrix}, \tag{3.22}$$

where J_a^b, J_r^b are given as in (3.17),(3.19). The error is

$$\bar{e} = \begin{bmatrix} e_a \\ e_r \end{bmatrix}. \tag{3.23}$$

The *absolute error* has a position and an orientation component and is given by

$$e_a = \begin{bmatrix} p_{ad}^b - p_a^b \\ \frac{1}{2}(S(n_a^b)n_{ad}^b + S(s_a^b)s_{ad}^b + S(a_a^b)a_{ad}^b) \end{bmatrix}, \tag{3.24}$$

where p_{ad}^b is the desired absolute position specified by the user in the base frame, p_a^b is the actual absolute position that can be computed as in (3.1), $n_{ad}^b, s_{ad}^b, a_{ad}^b$ are the column vectors of the rotation matrix R_{ad}^b giving the desired absolute orientation specified by the user in the base frame, and n_a^b, s_a^b, a_a^b are the column vectors of the rotation matrix R_a^b in (3.2). The *relative error* is given by

$$e_r = \begin{bmatrix} R_a^b p_{rd}^a - p_r^b \\ \frac{1}{2} R_1^b (S(n_r^1) n_{rd}^1 + S(s_r^1) s_{rd}^1 + S(a_r^1) a_{rd}^1) \end{bmatrix}. \qquad (3.25)$$

The rotation R_a^b is aimed at expressing the desired relative position p_{rd}^a, assigned by the user in the absolute frame, in the base frame; in this way, the specification of the desired relative position between the two end effectors is not affected by the absolute frame orientation. Further, in (3.25) notice that: p_r^b can be computed as in (3.3); $n_{rd}^1, s_{rd}^1, a_{rd}^1$ are the column vectors of the rotation matrix R_{rd}^1 giving the desired relative orientation specified by the user in the end-effector frame of the first manipulator; n_r^1, s_r^1, a_r^1 are the column vectors of the rotation matrix R_r^1 in (3.4); and the rotation R_1^b is aimed at expressing the orientation error in the base frame.

Finally, the desired velocity is

$$\bar{v}_d = \begin{bmatrix} v_{ad} \\ v_{rd} \end{bmatrix}. \qquad (3.26)$$

The *absolute velocity* term is given by

$$v_{ad} = \begin{bmatrix} \dot{p}_{ad}^b \\ \omega_{ad}^b \end{bmatrix}, \qquad (3.27)$$

where $\dot{p}_{ad}^b$ and ω_{ad}^b are respectively the desired absolute linear and angular velocities specified by the user in the base frame. The *relative velocity* term is given by

$$v_{rd} = \begin{bmatrix} R_a^b \dot{p}_{rd}^a + S(\omega_a^b) R_a^b p_{rd}^a \\ \omega_{rd}^1 \end{bmatrix}, \qquad (3.28)$$

where $\dot{p}_{rd}^a$ is the desired relative linear velocity specified by the user in the object frame and ω_{rd}^1 is the desired relative angular velocity specified by the user in the end-effector frame of the first manipulator. Notice that the expression of the translational part of the relative velocity presents an additional term which is a consequence of having assigned the relative position with reference to the absolute frame.

3.5 Cooperative system modeling

The dynamics of the two manipulators can be written in compact form as

$$M(q)\ddot{q} + C(q, \dot{q})\dot{q} + g(q) = \tau - J^{\mathrm{T}}(q)h, \qquad (3.29)$$

where the matrices are block-diagonal, e.g., $M = \text{blockdiag}\{M_1, M_2\}$, and the vectors are stacked, e.g., $g = [\,g_1^T \quad g_2^T\,]^T$. For each manipulator, τ_i is the vector of joint generalized forces, M_i is the symmetric positive-definite inertia matrix, $C_i\dot{q}_i$ is the vector of Coriolis and centrifugal generalized forces, g_i is the vector of gravitational generalized forces, and h_i is the vector of end-effector generalized forces.

The dynamics of the object is given by

$$M_o\dot{v}_o + C_o v_o + g_o = h_{ext}, \qquad (3.30)$$

where v_o is the vector expressing the (linear and angular) velocity of a frame Σ_o attached to the center of mass of the object, M_o is the object inertia matrix, $C_o v_o$ is the vector of Coriolis and centrifugal forces, g_o is the vector of gravitational forces, and h_{ext} is the vector of external forces acting at the center of mass of the object.

Under the assumption that the two manipulators tightly grasp a rigid object, holonomic constraints on both joint positions and velocities arise, e.g., see [13]. From a kinetostatic point of view, these constraints result in suitable mappings involving forces and velocities at the object level.

The mapping of the contact force vector h onto the external force vector h_{ext} is [4]

$$h_{ext} = \begin{bmatrix} I & O & I & O \\ S_1 & I & S_2 & I \end{bmatrix} h = Wh, \qquad (3.31)$$

where W is the grasp matrix, S_1 and S_2 are the matrices which transform the applied contact forces in moments at the object frame and depend on the grasp geometry, and I, O are the identity and null matrices of proper dimension, respectively.

The matrix W has full row rank; then, for a given h_{ext}, the inverse solution to (3.31) is given by

$$h = [\,W^\dagger \quad V\,] \begin{bmatrix} h_{ext} \\ h_{int} \end{bmatrix} = U^T \begin{bmatrix} h_{ext} \\ h_{int} \end{bmatrix}, \qquad (3.32)$$

where $W^\dagger$ denotes a pseudoinverse of W, V is a full-column-rank matrix spanning the null space of W, and the vector h_{int} represents the internal forces [14]. Notice that the following relation holds

$$WV = O. \qquad (3.33)$$

It has been recognized that the use in (3.32) of a generic pseudoinverse, e.g., the Moore-Penrose pseudoinverse, may lead to internal stresses even if

$h_{int} = 0$; to avoid this, $W^\dagger$ must be properly chosen [15]. In the remainder, it is assumed that the pseudoinverse of W has the following expression

$$W^\dagger = \begin{bmatrix} \frac{1}{2}I & O \\ -\frac{1}{2}S_1 & \frac{1}{2}I \\ \frac{1}{2}I & O \\ -\frac{1}{2}S_2 & \frac{1}{2}I \end{bmatrix}. \tag{3.34}$$

As pointed out in [15], this choice is also motivated by the work in [16] since it avoids problems with numeric solutions being noninvariant with respect to changes of scale or origin.

One possible choice for the matrix V is [14]

$$V = \begin{bmatrix} -I & O \\ S_1 & -I \\ I & O \\ -S_2 & I \end{bmatrix}. \tag{3.35}$$

In view of the duality between forces and velocities coming from the principle of virtual work, it can be recognized that

$$\begin{bmatrix} v_o \\ v_r \end{bmatrix} = \begin{bmatrix} W^{\dagger^\mathrm{T}} \\ V^\mathrm{T} \end{bmatrix} v, \tag{3.36}$$

where v_r is the relative velocity dual to h_{int} and v is the vector of end-effector velocities. Notice that tight grasp of a rigid object results in $v_r = 0$, i.e.,

$$V^\mathrm{T} J\dot{q} = 0. \tag{3.37}$$

3.6 Joint space control

The second stage of a kinematic control scheme for cooperative manipulators requires the development of a joint-space control law. In this case, it is assumed that reference joint trajectories performing a cooperative task are available through proper inverse kinematics.

First, a classical PD-type control law is considered; compensation of the gravity term is used in order to avoid steady-state errors. The control law for the system (3.29), (3.30), (3.32), is [17]

$$\tau = K_p\tilde{q} - K_d\dot{q} + g + J^\mathrm{T} W^\dagger g_o \tag{3.38}$$

where $\tilde{q} = q_d - q$ is the error between desired and actual joint positions; K_p and K_d are diagonal positive definite gain matrices of proper dimensions.

It has been shown in [17] that for a given set point q_d, the equilibrium ($\dot{q} = 0$, $\ddot{q} = 0$) of the system (3.29), (3.30), (3.32), under the control law (3.38) satisfies the condition

$$K_p\tilde{q}_{ss} - J^T V h_{int,ss} = 0 \qquad (3.39)$$

which reveals that at steady state an internal force is present if a joint error exists. Such an error may be due to inconsistency of the joint set point with the geometry of the grasp, that is, achievement of the joint set point would require violation of closed-chain constraints.

To avoid building of the internal force at steady state, a kineto-static filtering of the joint errors has been proposed which retains the simplicity of the above PD-type control law [18].

If the error term $K_p\tilde{q}$ is regarded as an elastic torque acting at the joints, it can be first transformed into the corresponding force at the two end effectors through J^{-T}, and then into an external force acting on the object through W. In this way, the part of the error term which builds internal force is filtered out. Thus, the control torque actually acting at the joints can be computed as the image of the required external force. The proposed control law is

$$\tau = J^T W^\dagger W J^{-T} K_p\tilde{q} - K_d\dot{q} + g + J^T W^\dagger g_o \qquad (3.40)$$

where it is assumed that the Jacobian matrix J is square ($n_1 + n_2 = 2m$) and full rank.

The equilibrium of the system (3.29), (3.30), (3.32), under the control law (3.40) satisfies

$$J^T W^\dagger W J^{-T} K_p\tilde{q}_{ss} - J^T V h_{int,ss} = 0. \qquad (3.41)$$

Since J^T is full column rank, it can be factored out and eq. (3.41) can be rewritten as

$$W^\dagger W J^{-T} K_p\tilde{q}_{ss} - V h_{int,ss} = 0. \qquad (3.42)$$

To analyze the equilibrium obtained, it is useful to observe that V spans the null space of W while $W^\dagger$ spans the range space of W^T; therefore, the two terms in the left-hand-side of (3.42) are orthogonal and thus they must be each equal to zero. Moreover, since $W^\dagger$ and V are full column rank matrices, it can be concluded that at steady state it is

$$\begin{cases} W J^{-T} K_p\tilde{q}_{ss} = 0 \\ h_{int,ss} = 0 \end{cases} \qquad (3.43)$$

Remarkably, the former condition implies that the component of the joint error term in the external-force space vanishes, the latter ensures that the internal force is null at steady state.

Therefore, the control law (3.40) cancels internal force at steady state, even if the joint set point cannot be reached due to closed-chain constraints. It is worth noting that the kineto-static filtering has no effect on steady-state errors due to external disturbances and thus the control law (3.40) reacts to them at full strength.

Equation (3.43) represents a set of constraints on the vector variable $\tilde{q}_{ss}$. The nonlinearity of the constraint equations does not allow drawing general conclusions about the solution of (3.43); physical reasoning, however, leads to conjecturing that a set of solution points exists, corresponding to different system equilibrium configurations.

3.7 Stability analysis

La Salle's global invariant set theorem —as reported in [19]— is invoked to analyze the stability of the equilibrium (3.43) [20-21].

Consider the scalar function with continuous first partial derivatives

$$V(x) = \frac{1}{2}\dot{q}^T M \dot{q} + \frac{1}{2}v_o^T M_o v_o + \frac{1}{2}\tilde{q}^T K_p \tilde{q}, \qquad (3.44)$$

where $x = [\tilde{q}^T \quad \dot{q}^T]^T$ belongs to the subspace $\Phi \subset \mathbb{R}^{24}$ constituted by the joint errors and velocities satisfying the closed-chain constraints. Notice that V in (3.44) is radially unbounded.

Under the assumption of a constant q_d (i.e., regulation problem), by using (3.29) and (3.30), the time derivative of (3.44) is

$$\dot{V}(x) = \dot{q}^T(\tau - J^T h - g) + v_o^T(h_{ext} - g_o) - \dot{q}^T K_p \tilde{q}, \qquad (3.45)$$

where the identities $\dot{q}^T(\dot{M} - 2C)\dot{q} = 0$, $v_o^T(\dot{M}_o - 2C_o)v_o = 0$ have been exploited.

By expressing v_o as in (3.36) and taking (3.32) into account, equation (3.45) can be rewritten as

$$\dot{V}(x) = \dot{q}^T(\tau - J^T V h_{int} - g - J^T W^\dagger g_o - K_p \tilde{q}). \qquad (3.46)$$

Considering the closed-chain constraint (3.37) and substituting control law (3.40) into (3.46) yields

$$\dot{V}(x) = -\dot{q}^T K_d \dot{q} - \dot{q}^T J^T (I - W^\dagger W) J^{-T} K_p \tilde{q}. \qquad (3.47)$$

It can be recognized that the second term on the right-hand side of (3.47) is null in view of the closed-chain constraint (3.37); in fact, the term $(I - W^\dagger W)J\dot{q}$ is a vector of end-effector velocities which correspond through (3.36) to sole relative velocities. At this point, equation (3.47) becomes

$$\dot{V}(x) = -\dot{q}^T K_d \dot{q} \tag{3.48}$$

which is negative semi-definite all over Φ.

The set R of all points $x \in \Phi$ where $\dot{V}(x) = 0$ is then

$$R = \{x \in \Phi : \quad \dot{q} = 0\}. \tag{3.49}$$

Following the analysis of the equilibrium in the previous Section, it can be recognized that the largest invariant set in R is

$$M = \{x \in R : \quad \tilde{q}_{ss} \text{ satisfies } (3.43)\}. \tag{3.50}$$

Therefore, the global invariant set theorem ensures global asymptotic convergence to M.

Notice that, according to the assumption on J being full-rank, the result is valid for any perturbation such that the resulting trajectory does not involve crossing of kinematic singularities of the two-manipulator system.

In the case when the set M contains the origin, i.e., when the given set point can be achieved without violating the closed-chain constraints, it is worth studying the domain of attraction of $x = 0$. Since $V(x)$ in (3.44) is quadratic, it is always possible to find a positive ℓ and a bounded region $\Phi_\ell \subset \Phi$ such that $\Phi_\ell \cap M = \{0\}$ and

$$\begin{cases} V(x) < \ell \\ \dot{V}(x) \leq 0 \end{cases} \quad \forall x \in \Phi_\ell. \tag{3.51}$$

By applying the local invariant set theorem, it can be recognized that Φ_ℓ is a domain of attraction for the equilibrium point $x = 0$.

When the set M does not contain the origin, the given set point cannot be achieved without violating the closed chain constraints. In this case, it becomes worth studying local stability of the minimum-norm element(s) in M; to the purpose, by following the same reasoning as above, the local invariant set theorem can be invoked to establish the existence of a domain of attraction.

3.7.1 Imperfect compensation of gravity terms

In many practical cases the mass of the object is not accurately known; thus, only a nominal estimate of the gravity term $\hat{g}_o$ is available. In this case the control law (3.40) becomes

$$\boldsymbol{\tau} = \boldsymbol{J}^{\mathrm{T}}\boldsymbol{W}^{\dagger}\boldsymbol{W}\boldsymbol{J}^{-\mathrm{T}}\boldsymbol{K}_p\tilde{\boldsymbol{q}} - \boldsymbol{K}_d\dot{\boldsymbol{q}} + \boldsymbol{g} + \boldsymbol{J}^{\mathrm{T}}\boldsymbol{W}^{\dagger}\hat{\boldsymbol{g}}_o. \qquad (3.52)$$

The equilibria of the system (3.29), (3.30) and (3.32) under the control law (3.52) satisfy

$$\begin{cases} \boldsymbol{W}\boldsymbol{J}^{-\mathrm{T}}\boldsymbol{K}_p\tilde{\boldsymbol{q}}_{ss} = \boldsymbol{g}_o - \hat{\boldsymbol{g}}_o \\ h_{int,ss} = 0 \end{cases} \qquad (3.53)$$

Inaccurate compensation of the object gravity term leads to a set of equilibrium configurations which are different from those obtained via (3.43). Nevertheless, the internal forces are still null at steady state.

Stability of the equilibrium (3.53) can be analyzed following the guidelines in [22]. To the purpose, consider the gravity energy functions $U_o(\boldsymbol{q})$ and $\hat{U}_o(\boldsymbol{q})$ such that

$$\frac{\partial U_o(\boldsymbol{q})}{\partial \boldsymbol{q}} = \boldsymbol{J}^{\mathrm{T}}\boldsymbol{W}^{\dagger}\boldsymbol{g}_o \qquad (3.54)$$

$$\frac{\partial \hat{U}_o(\boldsymbol{q})}{\partial \boldsymbol{q}} = \boldsymbol{J}^{\mathrm{T}}\boldsymbol{W}^{\dagger}\hat{\boldsymbol{g}}_o \qquad (3.55)$$

Notice that both $U_o(\boldsymbol{q})$ and $\hat{U}_o(\boldsymbol{q})$ are bounded for any $\boldsymbol{q}$.

At this point, consider the scalar function with continuous first partial derivatives

$$V_o(\boldsymbol{x}) = \frac{1}{2}\dot{\boldsymbol{q}}^{\mathrm{T}}\boldsymbol{M}\dot{\boldsymbol{q}} + \frac{1}{2}\boldsymbol{v}_o^{\mathrm{T}}\boldsymbol{M}_o\boldsymbol{v}_o + \frac{1}{2}\tilde{\boldsymbol{q}}^{\mathrm{T}}\boldsymbol{K}_p\tilde{\boldsymbol{q}} + U_o(\boldsymbol{q}) - \hat{U}_o(\boldsymbol{q}), \qquad (3.56)$$

which is obtained by suitably extending the scalar function V in (3.44). Again, $\boldsymbol{x}$ belongs to the subspace Φ constituted by the joint errors and velocities satisfying the closed-chain constraints. Notice that V_o is radially unbounded in view of radial unboundedness of V and boundedness of the gravity energy functions.

Taking into account (3.54,3.55), it can be easily recognized that

$$\dot{V}_o(\boldsymbol{x}) = -\dot{\boldsymbol{q}}^{\mathrm{T}}\boldsymbol{K}_d\dot{\boldsymbol{q}}, \qquad (3.57)$$

which is negative semi-definite all over Φ. Thus, the global invariant set theorem can be invoked to conclude asymptotic convergence of $\boldsymbol{x}$ to the invariant set

$$\mathrm{M}_o = \{\boldsymbol{x} \in \mathrm{R} : \tilde{\boldsymbol{q}}_{ss} \text{ satisfies } (3.53)\}. \qquad (3.58)$$

In the case of imperfect compensation of the manipulators' gravity term the same argument as above leads to recognize asymptotic stability of the equilibrium

$$\begin{cases} \boldsymbol{W}\boldsymbol{J}^{-\mathrm{T}}\boldsymbol{K}_p\tilde{\boldsymbol{q}}_{ss} = \boldsymbol{W}\boldsymbol{J}^{-\mathrm{T}}(\boldsymbol{g} - \hat{\boldsymbol{g}}) \\ \boldsymbol{V}h_{int,ss} = \boldsymbol{J}^{-\mathrm{T}}(\boldsymbol{g} - \hat{\boldsymbol{g}}) \end{cases} \qquad (3.59)$$

where $\hat{g}$ denotes the available estimate of g. It must be pointed out that in this case the equilibrium does not yield null internal force at steady state.

3.8 Addition of a force loop

The proposed control law (3.40) achieves the equilibrium with null internal force. If it is desired to impose a given internal force set point, force feedback should be added.

For a cooperative manipulator system, end-effector force measurements are typically available through wrist sensors. To extract internal forces from h, multiplication of (3.32) by $V^{\dagger}$ gives

$$h_{int} = V^{\dagger}h. \tag{3.60}$$

At this point, the control law (3.40) can be modified into

$$
\begin{aligned}
\tau \;=\; & J^{\mathrm{T}}W^{\dagger}WJ^{-\mathrm{T}}K_p\tilde{q} - K_d\dot{q} + g + J^{\mathrm{T}}W^{\dagger}g_o \\
& + J^{\mathrm{T}}V\big(h_{int,d} + K_f(h_{int,d} - V^{\dagger}h)\big),
\end{aligned}
\tag{3.61}
$$

where K_f is a positive definite matrix gain and $h_{int,d}$ is the desired internal force set point.

For a given set point q_d, the equilibrium of system (3.29), (3.30), and (3.32) under control law (3.61) satisfies

$$J^{\mathrm{T}}W^{\dagger}WJ^{-\mathrm{T}}K_p\tilde{q}_{ss} + J^{\mathrm{T}}V(I + K_f)(h_{int,d} - h_{int,ss}) = 0 \tag{3.62}$$

which can be rewritten as

$$
\begin{cases}
WJ^{-\mathrm{T}}K_p\tilde{q}_{ss} = 0 \\
h_{int,ss} = h_{int,d}.
\end{cases}
\tag{3.63}
$$

To analyze stability of equilibrium (3.63), consider the same scalar function $V(x)$ as in (3.44). It can be recognized that the time derivative of (3.44) along the system trajectory is the same as in (3.48); indeed, by substituting control law (3.61) into (3.46), the force feedback terms do not contribute to $\dot{V}(x)$ in view of the closed-chain constraint (3.37).

The same argument as in Section 3.7 leads to establishing global asymptotic convergence to the set

$$\mathrm{M}' = \{x \in \mathrm{R} : \quad \tilde{q}_{ss} \text{ satisfies } (3.63)\} \tag{3.64}$$

and local asymptotic stability of the minimum-norm element(s) in M'. Further, the same argument as in Subsection 3.7.1 can be used to analyze the case of imperfect compensation of the gravity terms.

3.9 Conclusions

A kinematic control approach for cooperative manipulator systems has been described. A useful feature of the chosen approach is that coordination is solved at inverse kinematics level while arm interactions are handled at joint control level. A set of meaningful variables has been defined to fully describe coordinated motion tasks in terms of absolute/relative position and orientation. An inverse kinematics algorithm has been presented whose outputs are the joint reference variables to be fed to joint-space controllers. In order to avoid building of internal forces kinetostatic filtering of the joint-space errors is adopted. Addition of a force control loop is also discussed.

Stability analysis has been provided for the equilibrium points of the cooperative system under the proposed joint-space control laws. The analysis has been extended to the case of imperfect compensation of the gravity terms.

Application of the above control strategy to a setup of two industrial manipulators is ongoing. Preliminary experimental results are reported in [23].

Acknowledgments

This work has been partly supported by Ministero dell'Università e della Ricerca Scientifica e Tecnologica under 60% and 40% funds.

References

[1] Y. Nakamura, K. Nagai, and T. Yoshikawa, "Mechanics of coordinative manipulation by multiple robotic mechanisms," *Proc. 1987 IEEE Int. Conf. on Robotics and Automation*, Raleigh, NC, 991–998, 1987.

[2] Z. Li, P. Hsu, and S. Sastry, "Grasping and coordinated manipulation by a multifingered robot hand," *Int. J. of Robotics Research*, 8(4), 33–50, 1989.

[3] P. Dauchez, *Descriptions de tâches en vue de la commande hybride symétrique d'un robot manipulateur a deux bras*, Thèse d'Etat, Université des Sciences et Techniques du Languedoc, Montpellier, F, 1990.

[4] M. Uchiyama and P. Dauchez, "Symmetric kinematic formulation and non-master/slave coordinated control of two-arm robots," *Advanced Robotics*, **7**, 361–383, 1993.

[5] P. Chiacchio, S. Chiaverini, and B. Siciliano, "Direct and inverse kinematics for coordinated motion tasks of a two-manipulator system," *Trans. ASME J. of Dynamic Systems, Measurement, and Control*, **118**, 691–697, 1996.

[6] L. Sciavicco and B. Siciliano, *Modeling and control of robot manipulators*, McGraw-Hill, New York, 1996.

[7] B. Siciliano, "A closed-loop inverse kinematic scheme for on-line joint-based robot control," *Robotica*, **8**, 231–243, 1990.

[8] P. Chiacchio, S. Chiaverini, L. Sciavicco, and B. Siciliano, "Closed-loop inverse kinematics schemes for constrained redundant manipulators with task space augmentation and task priority strategy," *Int. J. of Robotics Research*, **10**, 410–425, 1991.

[9] A. Balestrino, G. De Maria, L. Sciavicco, "Robust control of robotic manipulators," *Prep. 9th IFAC World Congress*, Budapest, H, **6**, 80–85, 1984.

[10] Y. Nakamura and H. Hanafusa, "Inverse kinematic solution with singularity robustness for robot manipulator control," *Trans. ASME J. of Dynamic Systems, Measurements, and Control*, **108**, 163–171, 1986.

[11] C.W. Wampler, "Manipulator inverse kinematic solutions based on vector formulations and damped least-squares methods," *IEEE Trans. on Systems, Man, and Cybernetics*, **16**, 93–101, 1986.

[12] Y. Nakamura, H. Hanafusa, and T. Yoshikawa, "Task-priority based redundancy control of robot manipulators," *Int. J. of Robotics Research*, **6**(2), 3–15, 1987.

[13] J.Y.S. Luh and Y.F. Zheng, "Constrained relations between two coordinated industrial robots for motion control," *Int. J. of Robotics Research*, **6**(3), 60–70, 1987.

[14] P. Chiacchio, S. Chiaverini, L. Sciavicco, and B. Siciliano, "Global task space manipulability ellipsoids for multiple arm systems," *IEEE Trans. on Robotics and Automation*, **7**, 678–685, 1991.

[15] I.D. Walker, R.A. Freeman, and S.I. Marcus, "Analysis of motion and internal loading of objects grasped by multiple cooperating manipulators," *Int. J. of Robotics Research*, **10**, 396-409, 1991.

[16] H. Lipkin and J. Duffy, "Hybrid twist and wrench control for a robotic manipulator," *Trans. ASME J. of Mechanisms, Transmissions, and Automation in Design*, **110**, 138–144, 1988.

[17] J.T. Wen and K. Kreutz, "Motion and force control of multiple robotic manipulators," *Automatica*, **28**, 729–743, 1992.

[18] P. Chiacchio and S. Chiaverini, "PD-type control schemes for cooperative manipulator systems," *Int. J. of Intelligent Automation and Soft Computing*, **2**, 65–72, 1996.

[19] J.-J.E. Slotine and W. Li, *Applied nonlinear control*, Prentice-Hall, Englewood Cliffs, NJ, 1991.

[20] F. Caccavale, P. Chiacchio and S. Chiaverini, "Stability analysis of a joint space control law for a two-manipulator system," *Proc. 35th Conf. on Decision and Control*, Kobe, J, 3008–3013, 1996.

[21] F. Caccavale, P. Chiacchio and S. Chiaverini, "Stability analysis of a joint space control law for a two-manipulator system," *IEEE Trans. on Automatic Control*, to appear, 1997.

[22] P. Tomei, "Adaptive PD controller for robot manipulators," *IEEE Transactions on Robotics and Automation*, vol. 7, no. 4, pp. 565–570, 1991.

[23] F. Caccavale, P. Chiacchio, S. Chiaverini, B. Siciliano, "Experiments of kinematic control on a two-robot system," *Prep. 11th CISM-IFToMM Symposium on theory and Practice of Robot Manipulators*, Udine, I, 1996.

Chapter 4

Load distribution and control of interacting manipulators

The chapter reviews a method for modeling and controlling two serial link manipulators which mutually lift and transport a rigid body object in a three dimensional workspace [31, 32, 33, 34]. A new vector variable is introduced which parameterizes the internal contact force controlled degrees of freedom. A technique for dynamically distributing the payload between the manipulators is suggested which yields a family of solutions for the contact forces and torques the manipulators impart to the object. A set of rigid body kinematic constraints which restricts the values of the joint velocities of both manipulators is derived. A rigid body dynamical model for the closed chain system is first developed in the joint space. The model is obtained by generalizing our previous methods for deriving the model. The joint velocity and acceleration variables in the model are expressed in terms of independent pseudovariables. The pseudospace model is transformed to obtain reduced order equations of motion and a separate set of equations governing the internal components of the contact forces and torques. A theoretic control architecture is suggested which explicitly decouples the two sets of equations comprising the model. The controller enables the designer to develop independent, non-interacting control laws for the position control and internal force control of the system.

4.1 Introduction

The problem of modeling and controlling two fixed base, serial link robotic manipulators to mutually lift and transport an object has been a subject of intensive study and research these past ten years. This interest has been motivated by the potential benefits of employing automatic and programmable two handed cooperative manipulation in diverse areas such as material handling and assembly. In the former application, two manipulators can cooperatively lift and transport large or voluminous objects that would be difficult or awkward for a single manipulator to move. Further, two cooperating manipulators can transport objects whose mass is beyond the lifting capacity of just one. Two cooperating manipulators can reduce the need for fixturing in many assembly applications, and may ultimately lead to fixtureless assembly in the air.

There have been numerous approaches proposed for modeling the interactions between the object and each manipulator and for controlling the forces and torques at the points of contact. In [1], models were developed which allow the contacts between the manipulators and object to be accidentally (e.g., due to slippage) or deliberately broken or the nature of the constraints changed due to wanted or unwanted disturbances. The analysis focused on a pair of two link planar revolute manipulators maintaining sliding point contacts with an object. The object was stabilized using a spring-dashpot combination.

In [2], it was proposed that a pair of six degree of freedom (DOF) manipulators maintain rolling point contacts with a rigid object. In the approach, three virtual revolute joints were added at the location of each effector. The kinematics of the rolling grasps were modeled.

The application of impedance control has resulted in successful implementations of two manipulators transporting an object [3, 4, 5]. These approaches enforce a controlled impedance of the manipulator endpoints or of the manipulated object itself.

This chapter, however, focuses on the case of two serial link manipulators mutually lifting and transporting objects that are rigid and jointless in a three dimensional workspace under the assumption of there being no relative motion between the end effectors and the object. That is to say, it is assumed that each manipulator securely holds the object without any slippage. The manipulators and object form a single closed chain mechanism, and there exists a large body of literature on modeling and controlling the manipulators in this configuration [6-33]. It should be mentioned that there have been some results reported for the case of two manipulators holding objects consisting of two rigid bodies connected by passive rotary or spherical joints [34, 35], where the assumption of no relative motion between each end effector and the rigid body it holds applied.

There are two challenging problems when modeling and controlling a dual manipulator closed chain system. First, the problem of dynamically distributing the load induced by the object between the manipulators is underspecified. Indeed, assuming that the object is rigid and jointless, its dynamical equations, i.e., Newton's and Euler's equations, are linear functions of the twelve components of contact force and torque the manipulators impart to it. Therefore, assuming that a reference trajectory for the center of mass of the object has been specified, there are infinitely many solutions for the contact forces and torques based on the object's dynamical equations. Each contact force[§1] solution contains a component that causes the object to move along the reference trajectory and a component that induces internal stress and torsion in the object but does not contribute to its motion. Various approaches for distributing the load have been proposed [7, 13, 17, 18, 19, 20, 26, 27, 28, 31].

The second problem is how to control the motion of the closed chain system and the contact forces. It has been shown that a set of six rigid body kinematic constraints are imposed on the values of the joint variables of both manipulators in this configuration [33]. Each constraint causes a loss of one position controlled DOF. This complicates the motion control problem because the number of actuated joints exceeds the number of positional DOF in the closed chain. If each manipulator is kinematically nonredundant, then the motion control objective is object trajectory tracking. If at least one of the manipulators is redundant, then there are additional positional DOF available to satisfy other objectives [36].

Another part of the control problem involves controlling or influencing the values of the internal component of the contact forces. Left unregulated, the internal forces could assume large values that result in the manipulators pulling against each other and would require large actuation torques at the joints while moving the object along its specified trajectory. Furthermore, excessively large values for the internal contact forces may even result in damage or deformity to the object or manipulators. There are two basic approaches to this problem: (i) to explicitly control the internal forces to track reference trajectories or (ii) to calculate the contact forces (including their internal components) by optimization techniques. In the explicit control case, some approaches proposed in the literature require knowledge of dynamics of the manipulators and object (e.g., see [10, 11]) while others do not (e.g. [9]). Most of the approaches that determine the contact forces to optimize a designer specified criteria involve no servoing and assume

---

[1]Contact force implies both contact force and contact torque hereinafter, unless otherwise specified.

knowledge of the dynamics of the held object [13, 17, 18, 19, 20, 31].

The chapter reviews our original approach for dynamic load distribution and explicit position- and internal force-control of the closed chain system consisting of two manipulators securely lifting and transporting a rigid body object in a three dimensional workspace [31, 32, 34]. The control architecture is dynamic model based, thus the chapter will also present a method for deriving a rigid body model for the system. The joint space model given here is a generalization of our previous techniques for modeling the system [32, 33]. It will be shown that the earlier results are just special cases of the modeling given here.

The chapter is organized as follows: A description of the system and the dynamical equations for the manipulators and object are given in Section 4.2. A general framework for load distribution is reviewed in Section 4.3. The kinematic coupling effects are modeled in Section 4.4 and a closed chain dynamical model in the joint space is derived in Section 4.5. A reduced order model governing the motion of the closed chain and a separate equation for calculating the internal components of the contact forces are the subject of Section 4.6. A control architecture originally proposed in [33] is reviewed in Section 4.7 where some recent insights into its net effect are discussed. A summary and conclusion are given in the final section.

4.2   System description and dynamics

The system is comprised of two serial link manipulators mutually holding and transporting a rigid body object in a three dimensional workspace. The manipulators and object form a single closed chain mechanism. Manipulator i ($i = 1, 2$) has a stationary base and contains N_i single DOF joints ($N_i \geq 6$ in the spatial case). The manipulators can be structurally distinct and possess different capabilities, i.e., they can have an equal ($N_1 = N_2$) or unequal ($N_1 \neq N_2$) number of joints. The object is rigid and jointless. It assumed that there is no relative motion between the end effectors and object, i.e., the end effectors securely hold the object without any slippage. The configuration of the system is shown in Figure 4.1.

4.2.1   System variables and coordinate frames

Let the joint positions, velocities, and accelerations of manipulator i be respectively represented by the ($N_i \times 1$) vectors $q_i = [q_{i1}, q_{i2}, \ldots, q_{iN_i}]^T$, $\dot{q}_i = [\dot{q}_{i1}, \dot{q}_{i2}, \ldots, \dot{q}_{iN_i}]^T$, and $\ddot{q}_i = [\ddot{q}_{i1}, \ddot{q}_{i2}, \ldots, \ddot{q}_{iN_i}]^T$. The joint positions of the two manipulators are the generalized coordinates describing

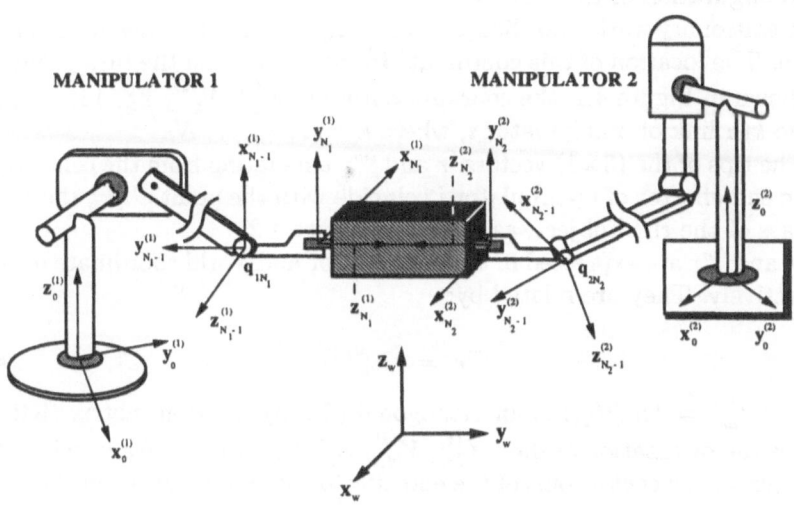

Figure 4.1: System configuration and coordinate system assignment

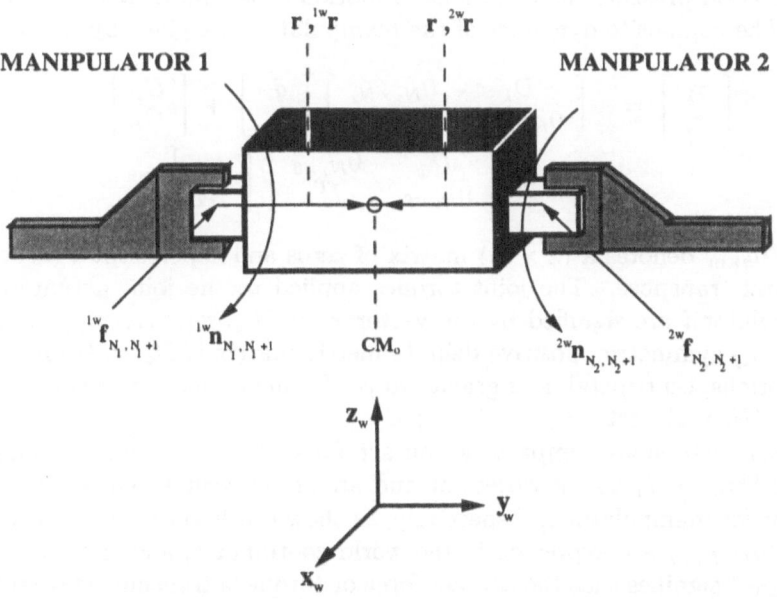

Figure 4.2: Freebody diagram for the common rigid object

the configuration of the system.

A stationary world coordinate frame (X_w, Y_w, Z_w) serves as a reference frame. The location of this coordinate frame is based on the task geometry. As shown in Figure 4.1, the coordinate frame ($X_k^{(i)}$, $Y_k^{(i)}$, $Z_k^{(i)}$) is assigned to the kth link of manipulator i, where $k = 1, 2, \ldots, N_i$.

The tips of the (3×1) vectors $^i r$ and $^{iw} r$ emanating from the centerpoint of the end effector of manipulator i coincide with the point CM_o, the center of mass of the rigid object, as shown in Figure 4.2.

$^i r$ and $^{iw} r$ are expressed in the end effector and world coordinate frames, respectively. They are related by:

$$^{iw} r = {}^i R_w^{N_i} \, {}^i r \tag{4.1}$$

where $^i R_w^{N_i} = {}^i R_w^{N_i}(q_i)$ is an orthogonal (3×3) rotation matrix that describes the orientation of the ($X_{N_i}^{(i)}$, $Y_{N_i}^{(i)}$, $Z_{N_i}^{(i)}$) coordinate frame which has its origin at the centerpoint of the end effector of manipulator i in the world coordinates.

4.2.2   Manipulator dynamics

This section presents the equations of motion of the individual manipulators. The composite dynamics of the manipulators are given by:

$$\begin{bmatrix} \tau_1 \\ \tau_2 \end{bmatrix} = \begin{bmatrix} D_1 & 0_{N_1 \times N_2} \\ 0_{N_2 \times N_1} & D_2 \end{bmatrix} \begin{bmatrix} \ddot{q}_1 \\ \ddot{q}_2 \end{bmatrix} + \begin{bmatrix} C_1 \\ C_2 \end{bmatrix}$$
$$+ \begin{bmatrix} J_{1w}^T & 0_{N_1 \times 6} \\ 0_{N_2 \times 6} & J_{2w}^T \end{bmatrix} \begin{bmatrix} f_{c1} \\ f_{c2} \end{bmatrix} \tag{4.2}$$

where $0_{k \times m}$ denotes a ($k \times m$) matrix of zeros and superscript T denotes a matrix transpose. The joint torques applied to the joint actuators of manipulator i are signified by the vector $\tau_i = [\tau_{i1}, \tau_{i2}, \ldots, \tau_{iN_i}]^T$. The ($N_i \times N_i$) symmetric, positive definite inertia matrix is $D_i = D_i(q_i)$, and the Coriolis, centripetal, and gravity forces for manipulator i are described by the ($N_i \times 1$) vector $C_i = C_i(q_i, \dot{q}_i)$.

Each manipulator imparts a contact force $^{iw} f_{N_i, N_i + 1}$ and a contact torque $^{iw} n_{N_i, N_i + 1}$ to the object at and about the centerpoint of the end effector for manipulator i, respectively, as shown in Figure 4.2. $^{iw} f_{N_i, N_i + 1}$ and $^{iw} n_{N_i, N_i + 1}$ are expressed in the world coordinates, and the subscript $N_i, N_i + 1$ signifies that the contact force or torque is transmitted from the N_ith link of manipulator i to the ($N_i + 1$)th link, where the latter link is the held object itself. The (6×1) vector f_{ci} in eq. (4.2) signifies the generalized contact force imparted by manipulator i. It is defined by:

$$f_{ci} = \begin{bmatrix} {}^{iw}f_{N_i,N_i+1} \\ {}^{iw}n_{N_i,N_i+1} \end{bmatrix} \qquad (4.3)$$

In eq. (4.2) , the $(N_i \times 6)$ transposed Jacobian matrix $J_{iw}^T = J_{iw}^T(q_i)$ transforms the generalized contact force[†2] imparted by manipulator i into the joint space. J_{iw} is assumed to possess full rank six.

4.2.3 Object dynamics

The dynamics for the rigid object are obtained through application Newton's and Euler's equations of motion. It is convenient to express these equations in a compact form:

$$Y = L \begin{bmatrix} f_{c1} \\ f_{c2} \end{bmatrix} \qquad (4.4)$$

In eq. (4.4) , Y is a (6×1) vector representing the net force (and torque) acting at the center of mass of the object due to its acceleration and gravity. It is defined by:

$$Y = \begin{bmatrix} m_c I_3 & 0_{3\times 3} \\ 0_{3\times 3} & K_c \end{bmatrix} \begin{bmatrix} \dot{v}_c \\ \dot{\omega}_c \end{bmatrix} + \begin{bmatrix} -m_c g \\ \Omega_c K_c \omega_c \end{bmatrix} = \Lambda \begin{bmatrix} \dot{v}_c \\ \dot{\omega}_c \end{bmatrix} + \begin{bmatrix} -m_c g \\ \Omega_c K_c \omega_c \end{bmatrix}$$
$$(4.5)$$

where I_k is a $(k \times k)$ identity matrix and where all Cartesian vectors are with respect to the world coordinate system (X_w, Y_w, Z_w). In eq. (4.5) , m_c is the mass of the rigid object, and K_c is the (3×3) symmetric inertia matrix of the object about its center of mass. The (3×1) vector g represents the gravitational acceleration of the object. The (6×1) vectors $[v_c^T, \omega_c^T]^T$ and $[\dot{v}_c^T, \dot{\omega}_c^T]^T$ denote the Cartesian velocity and acceleration of the center of mass of the object, respectively, with $(v_c, \dot{v}_c)$ being the translational and $(\omega_c, \dot{\omega}_c)$ the rotational components. The (6×6) matrix $\Lambda = \Lambda(m_c, K_c)$ is a compact representation of the coefficient matrix of $[\dot{v}_c^T, \dot{\omega}_c^T]^T$ in eq. (4.5)

In eq. (4.5) , $(\Omega_c K_c \omega_c)$ is a (3×1) vector arising from expressing the vector cross product $(\vec{\omega}_c \times (K_c \vec{\omega}_c))$ in a matrix-column vector notation, where Ω_c is a (3×3) skew symmetric matrix [33]:

$$\Omega_c = \begin{bmatrix} 0, & -\omega_{cz}, & \omega_{cy} \\ \omega_{cz}, & 0, & -\omega_{cx} \\ -\omega_{cy}, & \omega_{cx}, & 0 \end{bmatrix} \qquad (4.6)$$

---

[2]Generalized contact force will be referred to as contact force hereinafter.

and where $\omega_c = [\omega_{cx}, \omega_{cy}, \omega_{cz}]^T$.

The right side of eq. (4.4) represents the net force acting on the object at its center of mass due to the contact forces acting at the contact points between the manipulators and object. The (6×12) matrix L in eq. (4.4) is an explicit function of the (6×6) contact force transmission matrices L_1 and L_2 [33]:

$$L = [\ L_1, \quad L_2\ ] \tag{4.7}$$

where matrix $L_i (i = 1, 2)$ is defined by [33]:

$$L_i = \begin{bmatrix} I_3 & 0_{3\times3} \\ \begin{bmatrix} 0, & {}^{iw}r_z, & -{}^{iw}r_y \\ -{}^{iw}r_z, & 0, & {}^{iw}r_x \\ {}^{iw}r_y, & -{}^{iw}r_x, & 0 \end{bmatrix} & I_3 \end{bmatrix} = \begin{bmatrix} I_3 & 0_{3\times3} \\ E_i & I_3 \end{bmatrix} \tag{4.8}$$

In eq. (4.8), $E_i = E_i({}^{iw}r_x, {}^{iw}r_y, {}^{iw}r_z)$ is a (3×3) skew symmetric matrix arising from expressing the vector cross product $(-{}^{iw}\vec{r} \times {}^{iw}\vec{f}_{N_i, N_i+1})$ in a matrix-column vector notation, where $-{}^{iw}r$ represents a moment arm from point CM_o to point of application of f_{ci} (see Figure 4.2). It should be mentioned that $L_i = L_i(q_i)$ because ${}^{iw}r = {}^{iw}r(q_i)$ in accordance with eq. (4.1). Interestingly, eq. (4.8) reveals that L_i is nonsingular and that its determinant is equal to one.

In this chapter it is assumed that the joint variables of the manipulators in the closed chain configuration are known through feedback of their sensed or measured values or by feedback of their calculated values in a forward dynamic simulation of the system. Thus the nonlinear terms $\{D_i, C_i, J_{iw}\}$ in eq. (4.2) are known quantities. Furthermore, it is assumed that the object's mass, inertia, and geometric properties are known, and that a trajectory for the object's center of mass has been specified. Thus matrix L and vector Y in eq. (4.4) are known quantities.

4.3 A general framework for load distribution

To solve the underspecified dynamic load distribution problem, a new vector variable $\epsilon = [\epsilon_1, \epsilon_2, \ldots, \epsilon_6]^T$ is introduced. Six position controlled degrees of freedom (DOF) are lost due to the closed chain configuration [33]. The number of components of ϵ is equal to the dimension of the null space of matrix L and reflects the fact that the number of position controlled DOF lost is equal to the number of DOF gained for controlling the internal contact forces [18]. ϵ parameterizes the internal contact force DOF and is defined by:

$$\epsilon = M \begin{bmatrix} f_{c1} \\ f_{c2} \end{bmatrix} \tag{4.9}$$

The (6×12) matrix M in eq. (4.9) is selected such that the (12×12) composite matrix S, defined by:

$$S = \begin{bmatrix} L \\ M \end{bmatrix} \tag{4.10}$$

is nonsingular.

It is convenient to partition the inverse of S into two matrices:

$$S^{-1} = \begin{bmatrix} \Phi, & \Psi \end{bmatrix} \tag{4.11}$$

where Φ and Ψ are (12×6) matrices. Eqs. (4.10) and (4.11) imply five matrix identities:

$$L\Phi = I_6, \quad L\Psi = 0_{6 \times 6}, \quad M\Phi = 0_{6 \times 6}, \quad M\Psi = I_6, \quad \Phi L + \Psi M = I_{12} \tag{4.12}$$

where, here again, I_k and $0_{k \times l}$ denote a $(k \times k)$ identity matrix and a $(k \times l)$ matrix of zeros, respectively.

The identity $L\Psi = 0_{6 \times 6}$ reveals that the column vectors comprising Ψ lie in and span the null space of L. Observing eq. (4.7), an obvious choice for Ψ is:

$$\Psi = \begin{bmatrix} -L_1^{-1} \\ L_2^{-1} \end{bmatrix} \tag{4.13}$$

Matrix Ψ is not unique. Indeed, postmultiplying the choice for Ψ in eq. (4.13) by an arbitrary (6×6) nonsingular matrix yields a new Ψ which lies in the null space of L. In this chapter it is assumed that $\Psi = \Psi(L_1, L_2)$. Thus Ψ, like L, is a known quantity. The designer chooses M to satisfy $M\Psi = I_6$. Then, given $\{L, \Psi, M\}$, Φ is determined based on the matrix identities in eq. (4.12). These issues will be discussed later in this section.

Eqs. (4.4) and (4.9) can be solved for the contact forces [31, 32]:

$$\begin{bmatrix} f_{c1} \\ f_{c2} \end{bmatrix} = \Phi Y + \Psi \epsilon \tag{4.14}$$

in which eq. (4.11) has been invoked. The second term $\{\Psi\epsilon\}$ on the right of eq. (4.14) is the homogeneous solution to eq. (4.4) and is a component of $[f_{c1}^T, f_{c2}^T]^T$ which causes internal stress and torsion in the object but does not contribute to its motion since $L\Psi\epsilon = 0_{6\times1}$. The first term $\{\Phi Y\}$ on the right of eq. (4.14) is a particular solution to eq. (4.4) and is the component of $[f_{c1}^T, f_{c2}^T]^T$ which causes the object to physically move, since $L\Phi Y = Y$. However, it will be shown in this chapter that the particular solution to eq. (4.4) can contain a component which lies in the null space of L, and such a component causes internal stress and torsion in the object but does not contribute to its motion. This has been demonstrated previously in a dual manipulator context in [19] by a different approach which studied the characteristics of a class of pseudoinverses of L, but the approach given here is conceptually simpler.

The symbolic solution for the contact forces given by eq. (4.14) is significant because it indicates that the designer can specify the distribution of the payload's mass between the two manipulators by the choice of M and ϵ. For example, since Y is known, matrix Φ governs the distribution of the payload among the motion inducing components in the contact force solution.

4.3.1 Identifying motion inducing and internal stress components of (ΦY)

Any vector in the 12-dimensional linear space describing the contact forces imparted to the object by the manipulators can be expressed as linear combinations of two orthogonal subspaces: the exact range space of L^T and the null space Ψ of L. It is convenient to introduce the basis V:

$$V = [\ L^T,\ \ \Psi\ ] \tag{4.15}$$

It is easy to see that the columns vectors comprising V span the 12-dimensional linear space.

Matrix Φ can be expressed in terms of V:

$$\Phi = L^T\alpha + \Psi\gamma \tag{4.16}$$

where α and γ are (6×6) parameter matrices, respectively. It is easy to verify that $\alpha = (LL^T)^{-1}$ and $\gamma = -ML^T(LL^T)^{-1}$ by premultiplying eq. (4.16) by L and M, respectively, and noting eq. (4.12) . Substituting the solutions for $\{\alpha, \gamma\}$ into eq. (4.16) yields [31]:

$$\Phi = L^T(LL^T)^{-1} - \Psi M L^T(LL^T)^{-1} \tag{4.17}$$

Eq. (4.17) shows that $(\Phi\,Y)$ always contains a component $\{L^T\,(L\,L^T)^{-1}\,Y\}$ which contributes to the object's motion, but it may also contain a component $\{-\Psi\,M\,L^T\,(L\,L^T)^{-1}Y\}$ which induces internal stress and torsion in the object in the general case.

It is insightful to substitute for Φ in eq. (4.14) using eq. (4.17) :

$$\begin{bmatrix} f_{c1} \\ f_{c2} \end{bmatrix} = L^T\,(L\,L^T)^{-1}\,Y - \Psi\,\left(M\,L^T\,(L\,L^T)^{-1}\,Y - \epsilon\right) \qquad (4.18)$$

Eq. (4.18) describes all possible solutions to eq. (4.4) in terms of the basis V. Each solution in the family is distinguished by the designer's choice for the quantities $\{\Psi,\,M,\,\epsilon\}$ given $\{L,\,Y\}$. Interestingly, each and every distinct solution in the family has the identical object motion inducing component. Therefore the difference between any two distinct solutions lies in the null space of L.

4.3.2 Choosing matrix M

Matrix S is defined in eq. (4.10) . The purpose of this section is to determine a family of solutions for M which results in S being nonsingular and satisfies $M\,\Psi = I_6$ when Ψ is known. We then present three possible choices for M and calculate Φ for each of the choices. It is also shown how each choice for M can be obtained by selecting a parameter matrix in the family of solutions for M.

M can be expressed in terms of the basis V defined in eq. (4.15) :

$$M = \beta\,L + \zeta\,\Psi^T \qquad (4.19)$$

where β and ζ are (6×6) parameter matrices. It is easy to verify that $\zeta = (\Psi^T\,\Psi)^{-1}$ by postmultiplying eq. (4.19) by Ψ and observing eq. (4.12) . Substituting the solution for ζ in eq. (4.19) obtains:

$$M = \beta\,L + (\Psi^T\,\Psi)^{-1}\,\Psi^T \qquad (4.20)$$

When M is defined by eq. (4.20) , M^T will always contain a component that lies in the null space of L and therefore S will be nonsingular. Indeed, eq. (4.20) describes a family of solutions for M, and each distinct member of the family is characterized by the designer's choice for β.

Example 1: Choosing M to obtain a previous result

The dynamic load distribution problem that arises when two manipulators mutually lift a rigid object was not discussed in our earlier work [33] that modeled the closed chain configuration shown in Figure 4.1. The approach in [33] to modeling the dynamic coupling effects between the manipulators was to make the contact forces imparted by manipulator 1 implicit variables using the following procedure: (i) solve eq. (4.4) for $f_{c1}[= L_1^{-1}(Y - L_2 f_{c2})]$ (ii) substitute for f_{c1} into eq. (4.2) using its solution obtained in step i. The resulting equation represents the composite dynamics of both manipulators and the object and is an explicit function of f_{c2}. The physical interpretation of this modeling procedure was not discussed in [33].

In this example it is shown that the result of [33] can be obtained by an application of the general load distribution procedure presented here. The modeling procedure in [33] is obtained by selecting matrices Ψ and M to be:

$$\Psi = \begin{bmatrix} -L_1^{-1} L_2 \\ I_6 \end{bmatrix} \tag{4.21}$$

$$M = \begin{bmatrix} 0_{6\times 6}, & I_6 \end{bmatrix} \tag{4.22}$$

It should be noted that eq. (4.21) is obtained by postmultiplying the choice for Ψ in eq. (4.13) by L_2. Further, the choice for M in eq. (4.22) is obtained from eq. (4.20) by selecting β to be:

$$\beta = \left(\Psi^T \Psi\right)^{-1} L_2^T \left(L_1 L_1^T\right)^{-1} \tag{4.23}$$

Substituting eqs. (4.21) and (4.22) into eq. (4.17) yields the solution for Φ:

$$\Phi = \begin{bmatrix} L_1^{-1} \\ 0_{6\times 6} \end{bmatrix} \tag{4.24}$$

Substituting for $\{\Psi, \Phi\}$ in eq. (4.14) using eqs. (4.21) and (4.24) and inserting the result into eq. (4.2) yields the model in [33] where $\epsilon = f_{c2}$. The procedure in [33] has unknowingly distributed the load such that only manipulator 1 induces the object to physically move in space whereas the contact forces imparted by manipulator 2 are purely internal. In this extreme case, manipulator 1 bears the entire load.

Example 2: Choosing M to be a function of constrained parameters

Here Ψ is defined by eq. (4.13). In this example matrix M is selected to be a function of the force transmission matrices $\{L_1, L_2\}$ and two unknown scalar parameters $\{c_1, c_2\}$ whose values are restricted as follows [31, 32]:

$$c_1 + c_2 = 1 \tag{4.25}$$

Suppose M is chosen to be [31, 32]:

$$M = [-c_2 L_1, \quad c_1 L_2] \tag{4.26}$$

which is obtained from eq. (4.20) by selecting β to be:

$$\beta = c_1 I_6 - \left(\Psi^T \Psi \right)^{-1} \left(L_2 L_2^T \right)^{-1} \tag{4.27}$$

The symbolic solution for Φ can be determined by substituting for Ψ and M in eq. (4.17) using eqs. (4.13) and (4.26), respectively, and simplifying:

$$\Phi = \left[\begin{array}{c} c_1 L_1^{-1} \\ c_2 L_2^{-1} \end{array} \right] \tag{4.28}$$

The parameters $\{c_1, c_2\}$ will be treated as constants to be selected by the designer in the explicit internal force control approach given in this chapter. As an example, the solution for Φ given in eq. (4.24) is just a special case of eq. (4.28) with $\{c_1 = 1, c_2 = 0\}$. Alternatively, $\{c_1, c_2\}$ are viewed as variables when determining a solution for the internal contact forces by optimization techniques in [31].

It is repeated for emphasis that only the internal component of the particular solution (ΦY) to eq. (4.4) is a function of M. Therefore the terms in eq. (4.18) that are explicit functions of $\{c_1, c_2\}$ only affect the internal stress and torsion in the held object when eq. (4.26) applies.

Example 3: Choosing M so that M^T lies in the null space of L

This example is not dependent on a specific choice for matrix Ψ. Suppose that M is determined by choosing $\beta = 0_{6 \times 6}$ in eq. (4.20):

$$M = \left(\Psi^T \Psi \right)^{-1} \Psi^T \tag{4.29}$$

When eq. (4.29) applies, M^T lies in the null space of L, i.e., $L M^T = 0_{6 \times 6}$ and eq. (4.17) immediately simplifies:

$$\Phi = L^T \left(L L^T \right)^{-1} \tag{4.30}$$

Since the internal force component of (ΦY) has vanished, the terms (ΦY) and ($\Psi \epsilon$) in eq. (4.14) are now mutually orthogonal because:

$$\Phi^T \Psi = 0_{6 \times 6} \tag{4.31}$$

and orthogonality is the strongest form of linear independence between a pair of vectors [37].

The modeling of the kinematic coupling effects occurring between the manipulators is discussed next.

4.4 Modeling of kinematic coupling effects

There are two purposes for this section. First, a linear transformation relating the Cartesian velocity vector of the object and the vector of joint velocities for both manipulators will be derived. This relationship will be useful for expressing the object's dynamical equations in the joint space. Second, a set of rigid body kinematic constraints which must be satisfied by the joint velocities of the manipulators will be derived.

A linear relationship between the Cartesian velocity of the object at point CM_o and at the point of application of the contact force imparted by manipulator i, i.e., the centerpoint of the end effector, is established using the theory of infinitesimal rotation of a rigid object [38, 33]:

$$\begin{bmatrix} v_i \\ \omega_i \end{bmatrix} = L_i^T \begin{bmatrix} v_c \\ \omega_c \end{bmatrix} \tag{4.32}$$

where the (3×1) vectors v_i and ω_i represent the Cartesian translational and rotational velocities, respectively, of the end effector of manipulator i in the world coordinates.

Substituting for L_i^T in eq. (4.32) using eq. (4.8) verifies that $\omega_i = \omega_c$ as expected. Indeed, the Cartesian angular velocities of the end effectors and object are identical due to the assumption that the manipulators securely hold the object without any slippage.

Combining the two sets of equations obtained from eq. (4.32) with $i = 1, 2$ gives:

$$\begin{bmatrix} v_1 \\ \omega_1 \\ v_2 \\ \omega_2 \end{bmatrix} = \begin{bmatrix} L_1^T \\ L_2^T \end{bmatrix} \begin{bmatrix} v_c \\ \omega_c \end{bmatrix} = L^T \begin{bmatrix} v_c \\ \omega_c \end{bmatrix} \tag{4.33}$$

There is a well specified solution for the object velocities $[v_c^T, \omega_c^T]^T$ based on eq. (4.33) because L has full rank six and $[v_i^T, \omega_i^T]^T$ lies in the exact range space of L_i^T. The solution is obtained by premultiplying eq. (4.33) by matrix Φ^T and noting eq. (4.12) :

$$\begin{bmatrix} v_c \\ \omega_c \end{bmatrix} = \Phi^T \begin{bmatrix} v_1 \\ \omega_1 \\ v_2 \\ \omega_2 \end{bmatrix} \tag{4.34}$$

Three distinct solutions for Φ were obtained in the three examples of Section 4.3 given choices for Ψ and M. It is straightforward to verify that substituting for Φ^T in eq. (4.34) using each of the three solutions (for Φ) and applying eq. (4.32) yield $[v_c^T, \omega_c^T]^T = [v_c^T, \omega_c^T]^T$.

The velocities of the end effector of manipulator i in the Cartesian world coordinate frame and the joint space are related through the $(6 \times N_i)$ Jacobian matrix J_{iw}, i.e.:

$$\begin{bmatrix} v_i \\ \omega_i \end{bmatrix} = J_{iw}\,\dot{q}_i \tag{4.35}$$

Substituting for $[v_i^T, \omega_i^T]^T$ in eq. (4.34) using eq. (4.35) with $i = 1, 2$ relates the Cartesian velocities of the object at its center of mass to the joint space:

$$\begin{bmatrix} v_c \\ \omega_c \end{bmatrix} = \Phi^T \begin{bmatrix} J_{1w} & 0_{6 \times N_2} \\ 0_{6 \times N_1} & J_{2w} \end{bmatrix} \begin{bmatrix} \dot{q}_1 \\ \dot{q}_2 \end{bmatrix} = \Phi^T J \begin{bmatrix} \dot{q}_1 \\ \dot{q}_2 \end{bmatrix} \tag{4.36}$$

The $(12 \times (N_1 + N_2))$ composite Jacobian matrix $J = J(q_1, q_2)$ in eq. (4.36) has full rank twelve since it is assumed that J_{iw} has full rank six.

It is easy to see from Figure 4.1 that the end effectors of the manipulators cannot move independently when they mutually hold the rigid body object. The constraint between the Cartesian velocities of the end effectors is obtained by premultiplying eq. (4.33) by Ψ^T and noting eq. (4.12) :

$$\Psi^T \begin{bmatrix} v_1 \\ \omega_1 \\ v_2 \\ \omega_2 \end{bmatrix} = 0_{6 \times 1} \tag{4.37}$$

The constraint can be expressed in the joint space by substituting for $[v_i^T, \omega_i^T]^T$ in eq. (4.37) using eq. (4.35) with $i = 1, 2$:

$$\Psi^T J \begin{bmatrix} \dot{q}_1 \\ \dot{q}_2 \end{bmatrix} = A \begin{bmatrix} \dot{q}_1 \\ \dot{q}_2 \end{bmatrix} = 0_{6 \times 1} \tag{4.38}$$

where the $(6 \times (N_1 + N_2))$ matrix $A = A(q_1, q_2)\,(= \Psi^T J)$ is assumed to have full rank six.

Let $^k J$ denote the kth column vector of J, $(k = 1, 2, \ldots, N_1 + N_2)$. Since $^k J$ is a twelve dimensional vector, it can be expressed in terms of the basis V defined in eq. (4.15) :

$$^k J = L^T \alpha + \Psi \gamma \tag{4.39}$$

where α and γ are (6×1) parameter vectors. If $\gamma = 0_{6 \times 1}$ then the kth column of J lies in the null space of Ψ^T because $L\Psi = 0_{6 \times 6}$. It follows that the kth column of $A (= \Psi^{T\,k} J) = 0_{6 \times 1}$. In this case, none of the kinematic constraints in eq. (4.38) would be a function of the kth element of the vector of joint velocities $[q_1^T, q_2^T]^T$. Therefore it is further assumed that each column vector comprising J has a nonzero component lying in the null space of L.

Eq. (4.38) comprises six scalar constraint equations characterizing the kinematic dependence among the joint velocities when the manipulators operate in the closed chain configuration. Each independent scalar constraint contained in eq. (4.38) causes the loss of one position controlled DOF in the closed chain [38]. Indeed, the number of positional DOF in the entire closed chain system is $(N_1 + N_2 - 6)$. This is significant because the number of positional DOF specifies the number of independent ways that the dual-manipulator closed chain system can move without violating the constraints in eq. (4.38) .

A dynamical model for the multiple manipulator system in the joint space is presented next.

4.5 Derivation of rigid body model in joint space

The two manipulators and object form a single closed chain mechanism, and a rigid body model governing the motion of the closed chain and the behavior of the internal component of the contact forces is derived in the joint space in this section. In the ensuing development it is useful to define $N_{12} = N_1 + N_2$.

The first step in deriving this model is to substitute for $[f_{c1}^T, f_{c2}^T]^T$ in eq. (4.2) using eq. (4.14) :

$$\tau = \begin{bmatrix} D_1 & 0_{N_1 \times N_2} \\ 0_{N_2 \times N_1} & D_2 \end{bmatrix} \ddot{q} + \begin{bmatrix} C_1 \\ C_2 \end{bmatrix} + J^T \Phi Y + A^T \epsilon \qquad (4.40)$$

where J is defined in conjunction with eq. (4.36) and where $q = [q_1^T, q_2^T]^T$, $\dot{q} = [\dot{q}_1^T, \dot{q}_2^T]^T$, $\ddot{q} = [\ddot{q}_1^T, \ddot{q}_2^T]^T$, and $\tau = [\tau_1^T, \tau_2^T]^T$. Interestingly, it is observed that the coefficient matrix of ϵ in eq. (4.40) is just the transpose of the coefficient matrix of the vector of joint velocities in the kinematic constraints given by eq. (4.38) .

Vector Y in eq. (4.40) is a function of the Cartesian space variables $\{\omega_c, \dot{v}_c, \dot{\omega}_c\}$ according to its definition in eq. (4.5) . Y can be expressed in the joint space by substituting for ω_c and $[\dot{v}_c^T, \dot{\omega}_c^T]^T$ in eq. (4.5) using eq. (4.36) and its time derivative, respectively:

$$Y = \Lambda \Phi^T J \ddot{q} + \Lambda \left(\Phi^T \dot{J} + \dot{\Phi}^T J \right) \dot{q} + \left[\Omega_c K_c \left[0_{3\times3}, \quad I_3 \right] \Phi^T J \dot{q} \right]$$
$$\tag{4.41}$$

In eq. (4.41) , the (12×6) and $(12 \times N_{12})$ matrices $\dot{\Phi} [= (\partial\Phi/\partial q)\dot{q}]$ and $\dot{J} [= (\partial J/\partial q)\dot{q}]$, respectively, are both functions of the variables $\{q, \dot{q}\}$. The occurrence of ω_c on the right of eq. (4.5) has been replaced by $[0_{3\times3}, I_3] \Phi^T J \dot{q}$ in eq. (4.41) . The components $\{\omega_{cx}, \omega_{cy}, \omega_{cz}\}$ in matrix Ω_c are expressed in the joint space using this transformation, so $\Omega_c = \Omega_c(q, \dot{q})$ in eq. (4.41) .

Substituting for Y in eq. (4.40) using eq. (4.41) and rearranging terms yield the closed chain dynamics in the joint space:

$$\tau = D\ddot{q} + C + H_m\dot{q} + H_v + A^T \epsilon \tag{4.42}$$

The $(N_{12} \times N_{12})$ matrix $D = D(q)$ in eq. (4.42) is the inertia matrix for the entire system. It is defined by:

$$D = \begin{bmatrix} D_1 & 0_{N_1 \times N_2} \\ 0_{N_2 \times N_1} & D_2 \end{bmatrix} + J^T \Phi \Lambda \Phi^T J \tag{4.43}$$

Since D_i is positive definite, the first term to the right of eq. (4.43) is positive definite. The second term to the right of eq. (4.43) is positive semidefinite. Therefore D is positive definite because the sum of a positive definite matrix and a positive semidefinite matrix is positive definite [37]. The $(N_{12} \times 1)$ vector $C = C(q, \dot{q})$ is defined by:

$$C = \begin{bmatrix} C_1 \\ C_2 \end{bmatrix} \tag{4.44}$$

The $(N_{12} \times N_{12})$ matrix $H_m = H_m(q, \dot{q})$ and the $(N_{12} \times 1)$ vector $H_v = H_v(q, \dot{q})$ in eq. (4.42) are defined by:

$$H_m = J^T \Phi \Lambda \left(\Phi^T \dot{J} + \dot{\Phi}^T J \right) \tag{4.45}$$

$$H_v = J^T \Phi \left[\Omega_c K_c \left[0_{3\times3}, \quad -m_c g \atop I_3 \right] \Phi^T J \dot{q} \right] \tag{4.46}$$

It should be mentioned that the closed chain dynamical model derived in [32] is just a special case of eq. (4.42) with $\{\Psi, \Phi\}$ defined by eqs. (4.13) and (4.28) , respectively.

Eq. (4.42) accounts for the dynamics of all components of the closed chain but does not satisfy the rigid body kinematic constraints in eq. (4.38) . Indeed, eq. (4.42) , along with the time derivative of eq. (4.38) :

$$A\ddot{q} + \dot{A}\dot{q} = 0_{6\times1} \qquad (4.47)$$

comprise a joint space model which governs the motion of the closed chain dual manipulator system and the internal component of the contact forces. The $(6 \times N_{12})$ matrix $\dot{A}\,[= (\partial A/\partial q)\dot{q}]$ in eq. (4.47) is a function of the variables $\{q, \dot{q}\}$.

The form of eqs. (4.42) and (4.47) has been obtained for a broad class of constrained rigid body mechanical systems in [39, 40] using the method of Lagrange undetermined multipliers [38]. However, it is very unclear how the issues of dynamically distributing the load and relating ϵ to the internal contact forces would be addressed if the modeling techniques given in [39, 40] were applied to the multiple manipulator closed chain considered here.

To discuss the application of the joint space model to accomplish a forward dynamics simulation of the system, it is useful to combine eqs. (4.42) and (4.47) into a single equation:

$$\begin{bmatrix} D & A^T \\ A & 0_{6\times6} \end{bmatrix} \begin{bmatrix} \ddot{q} \\ \epsilon \end{bmatrix} = \begin{bmatrix} \tau - C - H_m\dot{q} - H_v \\ -\dot{A}\dot{q} \end{bmatrix} \qquad (4.48)$$

In the forward dynamics problem, the $(N_{12} + 6)$ quantities $\{\ddot{q}, \epsilon\}$ are unknowns and the joint torques τ are specified. A symbolic solution for the variables $\{\ddot{q}, \epsilon\}$ based on eq. (4.48) can be obtained by inverting the coefficient matrix of $[\ddot{q}^T, \epsilon^T]^T$ using inverse by partitioning [37]:

$$\ddot{q} = D^{-1}\Delta\,(\tau - C - H_m\dot{q} - H_v) - D^{-1}A^T\left(A D^{-1} A^T\right)^{-1}\dot{A}\dot{q} \quad (4.49)$$

$$\epsilon = \left(A D^{-1} A^T\right)^{-1}\left\{A D^{-1}\,(\tau - C - H_m\dot{q} - H_v) + \dot{A}\dot{q}\right\} \quad (4.50)$$

The solution for ϵ in eq. (4.50) is based on the invertibility of the quantity $(A D^{-1} A^T)$. D^{-1} is positive definite because D is. Given that A has full rank six, $(A D^{-1} A^T)$ is positive definite and therefore nonsingular. In eq. (4.49), Δ is a $(N_{12} \times N_{12})$ matrix defined by:

$$\Delta = I_{N_{12}} - A^T\left(A D^{-1} A^T\right)^{-1} A D^{-1} \qquad (4.51)$$

where, here again, $N_{12} = N_1 + N_2$ and $I_{N_{12}}$ signifies an $(N_{12} \times N_{12})$ identity matrix. By a mathematical observation, Δ is idempotent, i.e., $\Delta^2 = \Delta$, and therefore singular, since the only nonsingular idempotent matrix is the identity matrix [37]. It has been shown in our earlier work [33] that the rank of Δ equals the number of position controlled DOF in the closed chain, i.e., $rank\{\Delta\} = N_{12} - 6$.

While the joint space model is useful for understanding how the system evolves with time in response to applied joint torque inputs, it is not convenient for the controller design process. Indeed, the number of scalar equations in eq. (4.48) (or in eqs. (4.49) and (4.50) , which may also be viewed as a rigid body model) exceed the number of joint torque inputs. However, it is important to note that there is a well specified solution for τ based on the rigid body model. Since the rank of Δ equals $(N_{12} - 6)$ and D is positive definite, the rank of the coefficient matrix $(D^{-1}\Delta)$ of τ in eq. (4.49) is also equal to $(N_{12} - 6)$ [41]. Therefore an additional six independent scalar equations that are linear functions of τ are needed to yield a well specified solution for the N_{12} joint torques τ. The six equations are provided by eq. (4.50) . Rather than attempting to design a model based controller by solving eqs. (4.49) and (4.50) (or eq. (4.48)) for the joint torques, we will derive a reduced order model and design a control architecture based on it. This is discussed next.

4.6 Reduced order model

The joint velocities and accelerations form coupled sets of generalized velocities and accelerations for describing the configuration of the closed chain system, respectively. Linear transformations which express these variables in terms of new independent generalized velocities and accelerations are derived and then applied to eliminate $\{\dot{q}, \ddot{q}\}$ from the closed chain dynamical equations given by eq. (4.42) in this section. Then, building on the seminal work in [39], linear transformations are applied to eq. (4.42) to separate it into two sets of equations. The sets of equations govern the motion of the closed chain and the behavior of the internal component of the contact forces, respectively.

A new vector variable $\nu = [\nu_1, \nu_2, \ldots, \nu_{N_{12}-6}]^T$ referred to as the pseudovelocity vector [42, 43, 40] is introduced. The pseudovelocity vector is defined by:

$$\nu = B\dot{q} \tag{4.52}$$

where the $((N_{12}-6)\times N_{12})$ matrix $B = B(q)$ selected so that the composite $(N_{12} \times N_{12})$ matrix U, defined by:

$$U = \begin{bmatrix} A \\ B \end{bmatrix} \tag{4.53}$$

is nonsingular, where here again, A is defined in conjunction with eq. (4.38) and $N_{12} = N_1 + N_2$.

It is convenient to partition the inverse of U into two matrices:

$$U^{-1} = [\ \Upsilon, \ \Gamma \ ] \tag{4.54}$$

where $\Upsilon = \Upsilon(q)$ is an $(N_{12} \times 6)$ matrix and $\Gamma = \Gamma(q)$ an $(N_{12} \times (N_{12} - 6))$ matrix. Eqs. (4.53) and (4.54) imply five matrix identities:

$$\begin{aligned} A\,\Upsilon &= I_6 \\ A\,\Gamma &= 0_{6 \times (N_{12}-6)} \\ B\,\Upsilon &= 0_{(N_{12}-6) \times 6} \\ B\,\Gamma &= I_{N_{12}-6} \\ \Upsilon A + \Gamma B &= I_{N_{12}} \end{aligned} \tag{4.55}$$

The identity $A\,\Gamma = 0_{6 \times (N_{12}-6)}$ reveals that the column vectors comprising Γ lie in and span the null space of A. Γ can be determined by the following procedure. Noting that $A = \Psi^T J$ and $L\,\Psi = 0_{6 \times 6}$, six vectors lying in the null space (of A) are given by:

$$J^T \left(J\,J^T \right)^{-1} L^T$$

If $N_1 = N_2 = 6$, then the above set of vectors spans the null space and is assigned to Γ. If one or both of the manipulators is kinematically redundant, then $(N_{12} - 12)$ additional vectors are needed to span the null space. By a mathematical observation, $(N_{12} - 12)$ is the dimension of the null space of J, and any vector lying in the null space of J also lies in the null space of A. The null space of J can be determined by the zero eigenvalue matrix theorem [44].

All vectors lying in the N_{12}-dimensional articular space may be expressed in terms of the following basis Z:

$$Z = [\ A^T, \ \Gamma \ ] \tag{4.56}$$

It is straightforward to verify that Υ can be expressed in terms of this basis:

$$\Upsilon = A^T \left(A\,A^T \right)^{-1} - \Gamma B\,A^T \left(A\,A^T \right)^{-1} \tag{4.57}$$

Eqs. (4.38) and (4.52) can be solved for the joint velocities:

$$\dot{q} = \Gamma \nu \tag{4.58}$$

Differentiating eq. (4.52) with respect to time establishes the linear relationship between the pseudoaccelerations and the joint accelerations:

$$\dot{\nu} = B\ddot{q} + \dot{B}\dot{q} \tag{4.59}$$

The $((N_{12} - 6) \times N_{12})$ matrix $\dot{B}\,[= (\partial B/\partial q)\dot{q}]$ in eq. (4.59) is a function of the variables $\{q, \dot{q}\}$.

Eqs. (4.47) and (4.59) can be solved for $\ddot{q}$:

$$\ddot{q} = \Gamma\dot{\nu} - \left[\Upsilon\dot{A} + \Gamma\dot{B}\right]\Gamma\nu \tag{4.60}$$

where eq. (4.58) has been used. As a result, the matrices $\dot{A}\,[= (\partial A/\partial q)\Gamma\nu]$ and $\dot{B}\,[= (\partial B/\partial q)\Gamma\nu]$ in eq. (4.60) are now functions of $\{q, \nu\}$.

A solution for $\ddot{q}$ may also be obtained by differentiating eq. (4.58) with respect to time:

$$\ddot{q} = \Gamma\dot{\nu} + \dot{\Gamma}\nu \tag{4.61}$$

where the $(N_{12} \times (N_{12} - 6))$ matrix $\dot{\Gamma}[= (\partial\Gamma/\partial q)\Gamma\nu]$ is a function of the variables $\{q, \nu\}$.

Eqs. (4.60) and (4.61) are mathematically equivalent because of the following matrix identity:

$$\dot{\Gamma} = -\left[\Upsilon\dot{A} + \Gamma\dot{B}\right]\Gamma \tag{4.62}$$

Eq. (4.62) is obtained by differentiating the identity: $\Upsilon A + \Gamma B = I_{N_{12}}$ with respect to time and postmultiplying the resulting equation by Γ.

Substituting for $\dot{q}$ in eq. (4.38) using eq. (4.58) yields the kinematic constraint equation $A\Gamma\nu = 0_{6\times 1}$, which is identically true since $A\Gamma = 0_{6\times(N_{12}-6)}$. Therefore, the kinematic constraints at the velocity level are satisfied regardless of the values of the pseudovelocities when eq. (4.58) applies. Likewise, substituting for $\{\dot{q}, \ddot{q}\}$ in eq. (4.47) using eqs. (4.58) and (4.60) reveals that the kinematic constraints at the acceleration level are also satisfied regardless of the values of $\{\nu, \dot{\nu}\}$. These findings lead to the observation that expressing the closed chain dynamical model given by eqs. (4.42) and (4.47) in terms of the pseudovariables results in eq. (4.42) alone representing a rigid body model of the multiple manipulator system:

$$D\Gamma\dot{\nu} + A^T\epsilon = \tau - C - H_v + \left(D\left[\Upsilon\dot{A} + \Gamma\dot{B}\right] - H_m\right)\Gamma\nu \tag{4.63}$$

The number of equations in eq. (4.63) equals the sum of the position controlled DOF and the internal force controlled DOF in the closed chain system.

It is important to note that eq. (4.63) is still a nonlinear function of the joint positions q, i.e., $D = D(q)$, $C = C(q, \nu)$, $H_m = H_m(q, \nu)$, and $H_v = H_v(q, \nu)$. Thus it is difficult to perform a forward dynamics simulation of the system based on eq. (4.63) . However, as will now be shown, performing a linear transformation on eq. (4.63) makes the resulting set of equations valuable for controller design purposes.

Premultiplying eq. (4.63) by the nonsingular matrix $[\Gamma, D^{-1} A^T]^T$ and utilizing eq. (4.56) separates the model into two sets of equations governing the position controlled DOF and the internal force controlled DOF, respectively:

$$\Gamma^T D \Gamma \dot{\nu} = \Gamma^T \left\{ \tau - C - H_v + \left(D \left[\Upsilon \dot{A} + \Gamma \dot{B} \right] - H_m \right) \Gamma \nu \right\},$$
(4.64)

$$A D^{-1} A^T \epsilon = A D^{-1} \left\{ \tau - C - H_v - H_m \Gamma \nu \right\} + \dot{A} \Gamma \nu \qquad (4.65)$$

The $(N_{12} - 6)$ scalar equations comprising eq. (4.64) constitute the reduced order equations of motion for the closed chain system. Vector variable ϵ, which parameterizes the internal force controlled DOF, has been eliminated from eq. (4.64) which in turn is calculated as a function of the variables (q, ν, τ) using eq. (4.65) . Since D is positive definite and Γ and has full rank $(N_{12} - 6)$, then $(\Gamma^T D \Gamma)$ is positive definite and therefore nonsingular. $(A D^{-1} A^T)$ is positive definite and nonsingular by a similar argument given below eq. (4.50) . Thus eqs. (4.64) and (4.65) can be solved for $\dot{\nu}$ and ϵ, respectively.

Given the separated form of the reduced order model, we can now proceed with the controller design. This is discussed next.

4.7 Control architecture

The problem considered is to derive a control law for the N_{12} joint torques $\tau = [\tau_1^T, \tau_2^T]^T$ so that the variables $\{\epsilon, \nu\}$ quantifying the internal contact force- and position- controlled DOF can be controlled independently. This can be accomplished by applying the control architecture proposed in [33] to completely decouple eqs. (4.64) and (4.65) . The composite control $\{\tau\}$ is the sum of an $(N_{12} \times 1)$ primary controller τ^p and an $(N_{12} \times 1)$ secondary controller τ^s which are defined by:

$$\tau^p = - \left(D \left[\Upsilon \dot{A} + \Gamma \dot{B} \right] - H_m \right) \Gamma \nu + C + H_v,$$
(4.66)

$$\tau^s = A^T \tau_f^s + D \Gamma \tau_p^s$$
(4.67)

In eq. (4.67) , τ_f^s and τ_p^s are (6×1) and $((N_{12} - 6) \times 1)$ vectors, respectively, representing control variables to be determined.

The composite control $(\tau = \tau^p + \tau^s)$ defined by eqs. (4.66) and (4.67) is substituted into eqs. (4.64) and (4.65) . The resulting equations, under the assumption of perfect knowledge of the nonlinear terms in the model, leads to the closed loop system:

$$\dot{\nu} = \tau_p^s, \tag{4.68}$$

$$\epsilon = \tau_f^s \tag{4.69}$$

in which eq. (4.56) has been invoked. Derivation of eqs. (4.68) and (4.69) is based on the quantities $\{(\Gamma^T D \Gamma), (A D^{-1} A^T)\}$ being invertible. It was shown earlier that these quantities are positive definite and therefore non-singular.

Suppose τ_p^s is selected to servo the pseudovariable error, and τ_f^s for servoing the internal contact force error. Since eqs. (4.68) and (4.69) are completely decoupled, the secondary controller components τ_p^s and τ_f^s are non-interacting controllers for position and internal contact force, respectively.

It was claimed in [33] that the control architecture $\tau = \tau^p + \tau^s$ decoupled the control of the pseudovariables and an independent subset of the contact forces, namely those imparted by manipulator 2. As shown here in Example 1 of Section 4.3, the modeling procedure in [33] unknowingly distributed the load such that $\epsilon = f_{c2}$, i.e., the contact forces imparted by manipulator 2 are purely internal. The control law $(\tau = \tau^p + \tau^s)$ defined by eqs. (4.66) and (4.67) in fact decouples the position- and internal force-controlled DOF. The physical insight into the decoupling was first identified in [34]. It should be mentioned that a similar decoupling control architecture was developed independently by Wen et al. in [17].

4.8 Conclusions

The chapter has reviewed a method for modeling and controlling two serial link manipulators which mutually lift and transport a rigid body object in a three dimensional workspace. The system was viewed as a single closed chain mechanism and it was assumed that there is no relative motion between the end effectors and object. A new vector variable ϵ which parameterizes the internal contact force controlled degrees of freedom was introduced. It was defined as a linear function of the contact forces that both manipulators impart to the object using eq. (4.9) . A family of solutions to the dynamic load distribution problem was obtained by solving the object's dynamical equations and eq. (4.9) for the contact forces. The motion inducing component of every member of the family was shown to be identical. The internal component of the general load distribution solution was shown to contain two terms: $\{\Psi \epsilon\}$ and $\{- \Psi M L^T (L L^T)^{-1} Y\}$. Three choices for matrix M which transforms the contact forces to define ϵ in eq. (4.9) were suggested. Interestingly, the third choice caused the latter internal force term to vanish and resulted in the motion inducing and internal components of the solution being mutually orthogonal.

The kinematic coupling effects between the manipulators due to the shared payload were modeled. First, the Cartesian velocity of the object at its center of mass was expressed as a linear function of the joint velocities of both manipulators. Then a set of six rigid body kinematic constraints restricting the values of the joint velocities was derived.

A rigid body dynamical model for closed chain system consisting of $(N_1 + N_2 + 6)$ second order differential equations was first derived in the joint space. The upper $(N_1 + N_2)$ equations in the model are the closed chain dynamical equations. They were derived by substituting the load distribution solution for the contact forces into the manipulators' dynamical equations. The resulting equations are linear functions of the Cartesian vector Y defined in eq. (4.5) . We proposed here a generalization of our previous methods [32, 33] for expressing Y in the joint space where $Y = Y(q, \dot{q}, \ddot{q})$ becomes an explicit function of the matrix Φ. Our previous results can be obtained by specifying choices for Φ in eq. (4.41) .

The last six equations in the joint space model are the kinematic acceleration constraints. By expressing the model in the pseudospace, it was shown that these last six equations are satisfied regardless of the values of the pseudovariables. Therefore the upper $(N_1 + N_2)$ equations of the model, when expressed in the pseudospace, comprise a rigid body model for the system. Linear transformations were applied to the $(N_1 + N_2)$ equations in the model to obtain reduced order equations governing the motion of the system and a separate set of equations governing the internal components of the contact forces. Both sets are functions of the joint torques of both manipulators, but only the latter is a function of ϵ. The control architecture originally proposed in [33] was applied to completely decouple the two sets of equations comprising the separated form of the model. As a result, the pseudovariables and the elements of ϵ are controlled independently.

Acknowledgements

This research was sponsored by the Office of Engineering Research Program, Basic Energy Sciences, U.S. Department of Energy, under Contract No. DE-

AC05-96OR22464 with Lockheed Martin Energy Research Corporation.
The author wishes to thank Dr. Lynne E. Parker for encouraging the continuation and completion of this research.

References

[1] K. Laroussi, H. Hemami, and R.E. Goddard, "Coordination of Two Planar Robots in Lifting," *IEEE Journal of Robotics and Automation,* vol. 4, no. 1, pp. 77-85, February 1988.

[2] P. Chiacchio, S. Chiaverini, and B. Siciliano, "Direct and inverse kinematics for coordinated motion tasks of a two-manipulator system," *Trans. ASME J. of Dynamic Systems, Measurement, and Control,* vol. 118, pp. 691–697, December 1996.

[3] R. Bonitz and T. Hsia, "Robust Internal Force-tracking Impedance Control for Coordinated Multi-arm Manipulation - Theory and Experiments" *Robotic and Manufacturing Systems,* (Proc. World Automation Congress (WAC'96), May 28-30, 1996, Montpellier, France) edited by M. Jamshidi and F.G. Pin; TSI Press Series, pp. 111-118, 1996.

[4] R. Bonitz and T. Hsia, "The Effects of Computational Delays in Coordinated Multiple-arm Manipulation Using Robust Internal Force-based Impedance Control" *Robotic and Manufacturing Systems,* (Proc. World Automation Congress (WAC'96), May 28-30, 1996, Montpellier, France) edited by M. Jamshidi and F.G. Pin; TSI Press Series, pp. 103-110, 1996.

[5] S. Schneider and R. Cannon, "Object Impedance Control for Cooperative Manipulation: Theory and Experimental Results" *IEEE Trans. Robotics and Automation,* vol. 8, no. 3, pp. 383-394, 1992.

[6] M. Uchiyama, T. Kitano, Y. Tanno, and K. Miyawaki, "Cooperative Multiple Robots to be Applied to Industries" *Robotic and Manufacturing Systems,* (Proc. World Automation Congress (WAC'96), May 28-30, 1996, Montpellier, France) edited by M. Jamshidi and F.G. Pin; TSI Press Series, vol. 3, pp. 759-764, 1996.

[7] M. Uchiyama, X. Delebarre, H. Amada, and T. Kitano, "Optimum Internal Force Control for Two Cooperative Robots to Carry an Object", *Intelligent Automation and Soft Computing,* (Proc. World Automation Congress (WAC'94), Maui, HI, August 14-17, 1994) vol. 2, pp. 111-116, TSI Press Series, 1994.

[8] M. Uchiyama and P. Dauchez, "Symmetric kinematic formulation and non-master slave coordinated control of two-arm robots", *Advanced Robotics*, vol. 7, no. 4, pp. 361-383, 1993.

[9] P. Chiacchio and S. Chiaverini, "PD-Type Control Schemes For Cooperative Manipulator Systems," *Intelligent Automation and Soft Computing,* (journal) vol. 2, no. 1, pp. 65-72, 1996.

[10] O. Khatib, K. Yokai, K. Chang, D. Ruspini, R. Holmberg, and A. Casal, "Cooperative Tasks in Multiple Mobile Manipulation Systems," *Robotics and Manufacturing*, (Proc. Sixth Int'l Symposium on Robotics and Manufacturing (ISRAM'96), May 28-30, 1996, Montpellier, France) edited by M. Jamshidi et. al., vol. 6, pp. 345-350, ASME Press, 1996.

[11] D. Williams and O. Khatib, "Modeling and Control of Internal Force Dynamics in Multi-Grasp Manipulation", *Robotics and Manufacturing*, (Proc. Fifth Int'l Symposium on Robotics and Manufacturing (ISRAM'94), Maui, HI, August 14-18, 1994) edited by M. Jamshidi et. al., vol. 5, pp. 735-741, ASME Press, 1994.

[12] P. Hsu, "Adaptive Internal Force Control of a Two Manipulator System," *Robotics and Manufacturing*, (Proc. Fifth Int'l Symposium on Robotics and Manufacturing (ISRAM'94), Maui, HI, August 14-18, 1994) edited by M. Jamshidi et. al., vol. 5, pp. 151-156, ASME Press, 1994.

[13] P. Hsu, "Coordinated Control of Multiple Manipulator Systems", *IEEE Trans. Robotics and Automation*, vol. 9, no. 4, pp. 400-410, August, 1993.

[14] D.J. Cox and D. Tesar, "Development System Environment For Dual-Arm Robotic Operations," *Robotics and Manufacturing*, (Proc. Fifth Int'l Symposium on Robotics and Manufacturing (ISRAM'94), Maui, HI, August 14-18, 1994) edited by M. Jamshidi et. al., vol. 5, pp. 61-66, ASME Press, 1994.

[15] D.J. Cox and D. Tesar, "Cooperative Dual-Arm Robotic Operations with Fixture Interaction," *Intelligent Automation and Soft Computing,* (Proc. World Automation Congress (WAC'94), Maui, HI, August 14-17, 1994) vol. 2, pp. 439-444, TSI Press Series, 1994.

[16] F. Caccavale and J. Szewczyk, "Experimental Results of Operational Space Control on a Dual-Arm Robot System," *Robotics and Manufacturing*, (Proc. Sixth Int'l Symposium on Robotics and

Manufacturing (ISRAM'96), May 28-30, 1996, Montpellier, France) edited by M. Jamshidi et. al., vol. 6, pp. 121-126, ASME Press, 1996.

[17] J.T. Wen and K. Kreutz Delgado, "Motion and Force Control of Robotic Manipulators," *Automatica*, vol. 28, pp. 729-743, 1992.

[18] K. Kreutz and A. Lokshin, "Load Balancing and Closed Chain Multiple Arm Control," *American Control Conf.*, Atlanta, GA, vol. 3, pp. 2148-2155, June 1988.

[19] I.D. Walker, R.A. Freeman, and S.I. Marcus, "Analysis of Motion and Internal Loading of Objects Grasped by Multiple Cooperating Manipulators," *Int'l J. of Robotics Research*, vol. 10, no. 4, pp. 396-409, August 1991.

[20] I.D. Walker, S.I. Marcus, and R.A. Freeman, "Distribution of Dynamic Loads for Multiple Cooperating Robot Manipulators," *J. Robotic Systems*, vol. 6, no. 1, pp. 35-47, January 1989.

[21] F. Pfeiffer, "Cooperating Fingers - A Special Form of Cooperating Robots," *Robotic and Manufacturing Systems*, (Proc. World Automation Congress (WAC'96), May 28-30, 1996, Montpellier, France) edited by M. Jamshidi and F.G. Pin, pp. 639-643, TSI Press Series, 1996.

[22] W. Nguyen and J.K. Mills "Fixtureless Assembly: Multi-Robot Manipulation of Flexible Payloads," *Robotics and Manufacturing*, (Proc. Sixth Int'l Symposium on Robotics and Manufacturing (ISRAM'96), May 28-30, 1996, Montpellier, France) edited by M. Jamshidi et. al., vol. 6, pp. 661-666, ASME Press, 1996.

[23] M.E. Pittelkau, "Adaptive Load Sharing Force Control for Two-Arm Manipulators," *IEEE Int'l Conf. Robotics and Automation*, vol. 1, pp. 498-503, Philadelphia, PA, April 24-29, 1988.

[24] Y.-R. Hu and A.A. Goldenberg, "An Adaptive Approach to Motion and Force Control of Multiple Coordinated Robot Arms," *IEEE Int'l Conf. Robotics and Automation*, vol. 2, pp. 1091-1096, Scottsdale, AZ, May 14-19, 1989.

[25] Y. Nakamura, K. Nagai, and T. Yoshikawa, "Dynamics and Stability in Coordination of Multiple Robotic Mechanisms," *Int'l J. of Robotics Research*, Vol. 8, No. 2, pp. 44-61, April 1989.

[26] Y.D. Shin and M.J. Chung, "An Optimal Force Distribution Scheme for Cooperating Multiple Robot Manipulators", *Robotica*, pp. 49-60, Vol. 11, Part 1, Jan.-Feb. 1993.

[27] M.H. Choi, B.H. Lee, and M.S. Ko, "Optimal Load Distribution for Two Cooperating Manipulators Using a Force Ellipsoid", *Robotica*, pp. 61-72, Vol. 11, Part 1, Jan.-Feb. 1993.

[28] L.T. Wang and M.J. Kuo, "Dynamic Load Carrying Capacity and Inverse Dynamics of Multiple Cooperating Manipulators", *IEEE Trans. Robotics and Automation*, pp. 71-77, Vol. 10, no. 1, February 1994.

[29] X. Yun, "Modeling and Control of Two Constrained Manipulators", *Journal of Intelligent and Robotic Systems*, pp. 363-377, Vol. 4, 1991.

[30] X. Yun and V.R. Kumar, "An Approach to Simultaneous Control of Trajectory and Interaction Forces in Dual-Arm Configurations", *IEEE Trans. on Robotics and Automation*, pp. 618-625, Vol. 7, No. 5, Oct. 1991.

[31] M.A. Unseren, "Determination of Contact Forces for Two Manipulators Mutually Lifting a Rigid Object Using a Technique of Dynamic Load Distribution, " *Intelligent Automation and Soft Computing* (journal), Vol. 2, No. 1, pp. 49-63, 1996.

[32] M.A. Unseren, "A New Technique for Dynamic Load Distribution when Two Manipulators Mutually Lift a Rigid Object. Part 1: The Proposed Technique (pp. 359-365); Part 2: Derivation of Entire System Model and Control Architecture (pp. 367-372)"; *Intelligent Automation and Soft Computing, Trends in Research, Development, and Applications*, (Proc. World Automation Congress (WAC '94), August 14-17, 1994, Maui, HI) edited by M. Jamshidi, etc.; TSI Press Series, 1994.

[33] M.A. Unseren, "Rigid Body Dynamics and Decoupled Control Architecture for Two Strongly Interacting Manipulators," *Robotica*, vol. 9, part 4, pp. 421-430, 1991.

press),

[34] M.A. Unseren, "A Rigid Body Model and Decoupled Control Architecture For Two Manipulators Holding a Complex Object," *Robotics and Autonomous Systems*, Vol. 10, No. 2-3, pp. 115-131, 1992.

[35] J.Y.S. Luh and Y.F. Zheng, "Constrained Relations Between Two Coordinated Industrial Robots For Motion Control," *International Journal of Robotics Research,* vol. 6, no. 3, pp. 60-70, Fall 1987.

[36] M.A. Unseren, "An Approach to Modeling a Kinematically Redundant Closed Chain System Using Pseudovelocities," *Robotics and Manufacturing,* (Proc. Sixth Int'l Symposium on Robotics and Manufacturing (ISRAM'96), May 28-30, 1996, Montpellier, France) edited by M. Jamshidi et. al., vol. 6, pp. 843-850, ASME Press, 1996.

[37] B. Noble and J.W. Daniel, *Applied Linear Algebra,* 2nd ed., Prentice-Hall, Inc., 1977.

[38] H. Goldstein, *Classical Mechanics,* 2nd edition, Addison-Wesley, 1980.

[39] H. Hemami and F.C. Weimer, "Modeling of Nonholonomic Dynamic Systems with Applications," *ASME J. of Applied Mechanics,* vol. 48, pp. 177-182, March 1981.

[40] R.K. Kankaanranta and H.N. Koivo, "Dynamics and Simulation of Compliant Motion of a Manipulator," *IEEE J. Robotics and Automation,* vol. 4, no. 2, pp. 163-173, April 1988.

[41] D.T. Finkbeiner, *Introduction to Matrices and Linear Transformations,* 3rd ed., W.H. Freeman and Company, 1978.

[42] F.R. Gantmacher, *Lectures in Analytical Mechanics,* USSR: Mir Publishers, chap. 1, sec. 10, 1975.

[43] H. Hemami, "A Feedback On-Off Model of Biped Dynamics," *IEEE Trans. Systems, Man, and Cybernetics,* vol. SMC-10, no. 7, pp. 376-383, sec. IV, July 1980.

[44] J.W. Kamman and R.L. Huston, "Dynamics of Constrained Multibody Systems," ASME Journal of Applied Mechanics, Vol. 51, pp. 899-903, December 1984.

Chapter 5

Multi-fingered hands:
A survey

Humans have evolved as the dominant species on the planet in part because of their skills in fine manipulation using their multifingered hands. In recent years, there has been much activity in the design, analysis, and control of artificial multi-fingered hands, and corresponding research in the area of machine dexterity. This chapter presents a brief survey of these efforts, and attempts to provide an extensive bibliography of the area.

5.1   Robot hand hardware

We begin with a brief review of the state of the art in robot hand hardware. In the last fifteen years, significant progress has been made in the development of dextrous robot end effectors, which previously were largely constrained to variants of the parallel jaw gripper. Some specialized single degree of freedom grippers have been successfully introduced. However, these are largely limited to the case where the objects to be grasped are in a small, well-understood set, and are not truly dextrous in the general sense.

Some early three-fingered hands, such as the Jameson Hands [47] began the trend of development of more sophisticated end effectors. Numerous multifingered hands have since been built and successfully demonstrated, notably the Salisbury hand [81, 102, 111] (Figure 5.1) and the MIT/Utah hand [55, 109, 116] (Figure 5.2).

These two hands have become the standards for researchers involved in robot hand algorithm development and laboratory experimentation, partic-

Figure 5.1: Salisbury Hand photograph (courtesy of NASA).

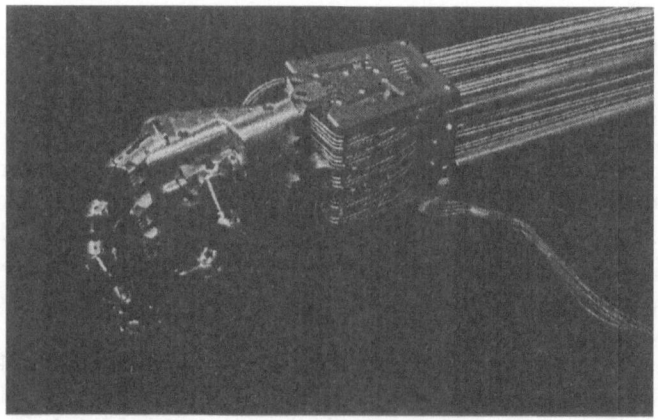

Figure 5.2: MIT/Utah Hand photograph (courtesy of NASA).

ularly in the USA. Other hands of note include the Darmstadt Hand [13], the Karlsruhe hand [134], the Bologna Hand [82], the Anthrobot Hand [2], the Belgrade-USC Hand [105] and the Waseda series of hands [63]. Design and analysis of new hands continues [35, 36, 77, 120].

A wide range of design strategies have been followed in the production of these hands. There are three [13, 77, 81, 82, 120], four [55], and five-fingered [2, 36, 63] hands. A number of different arrangements fo the fingers have been adopted, although the most popular arrangement by far mimics that of the human hand, with a 'thumb' opposing two or more 'fingers'. Some hands are tendon-driven [2, 13, 55, 81, 82], and some powered by actuators in the hand unit itself [36, 77, 120]. Electric motors [2, 77, 81, 82, 120], hydraulics [55], and pneumatic [55] power units have been employed as actuation devices.

Numerous different types of sensors have been suggested and implemented on robot hands. For finger control, in addition to joint position sensors (encoders, potentiometers, Hall Effect sensors, etc.), a common remote sensing mode has been that of force sensing via strain gages [13]. In some cases the strain gages are installed directly at the fingers themselves [77, 120], and in other cases they have been mounted remote from the hand, sensing forces via tendon tensions [81].

For environmental sensing and measurement, various contact and noncontact sensors have been proposed. These range from resistive and capacitive fingertip sensors or sensor arrays [42] to infrared and other proximity sensors at the finger joints and elsewhere [77]. Vision has also been successfully used [77]. Contact sensing is a particularly difficult issue, since our intuition about hand sensors is based on the existence of a rich, dense, and highly varied set of sensors embedded in the skin of human hands [42]. Although tactile sensing technology is improving rapidly [7], it will be a long time before robot hands can rival human hands for sensor quantity and variety. This lack of sensor richness has proved an obstacle to robot hand development, however, numerous creative solutions are being developed.

With the exception of the Barrett hand [120], which has been designed specifically for industrial applications, and possibly the Belgrade-USC Hand [105], most of the above hands have been confined to the laboratory at the time of writing, and this trend is likely to continue. There are numerous reasons for this. Many of the hands (including the Salisbury and MIT/Utah hands) were designed to be research testbeds, supporting theoretical and algorithmic research rather than being immediately practical devices. In addition, many of the current generation of hands have bulky remote actuation packages (the Barrett Hand and the recent self-contained Hirzinger hand [35] are notable exceptions) which make transition to applications difficult. Reliability, control interfaces, and a lack of good sensory capabilities are also issues of concern.

However, another key obstacle, which we will concentrate on in the following, has been the sheer complexity involved in modeling and control of dextrous multifingered tasks. Although the efficient use of multifingered hands is familiar to almost all humans, the understanding and translation of this skill to robot hands is a significant and fascinating problem. In the following section, we will review some of the issues which make multifingered manipulation a unique undertaking. The remaining sections in this chapter attempt to provide a brief summary of the efforts researchers have made to address these issues to date.

5.2   Key issues underlying multifingered manipulation

Given a particular robot hand, the kinematic and dynamic (if desired) models of each finger can be readily obtained using techniques previously established for robot manipulators. However taking the next step, and modeling dextrous multifingered manipulation itself, is not an trivial undertaking. The essential difficulty is in modeling the interaction between the fingers and the object.

Successful multifingered grasping can be viewed as an extension of the case of cooperation among multiple manipulator arms. The essential difference lies in the nature of the contacts between the manipulators (fingers) and the grasped object. For the case of cooperating robot arms, where each arm has a solid grasp of the object, there is an extensive body of literature [22, 72, 93, 97, 119, 127, 128, 132, 143]. and modeling of the situation is fairly well understood [28, 46, 65, 67, 71, 131, 133, 135].

However, for the case of multi-fingered manipulation, the situation is complicated by the fact that the fingertips are not solidly attached to the held object, as in the typical multi-arm coordination problem. The whole essence of dextrous multifingered manipulation lies in the ability of the fingertips to move relative to a held object. This causes extra complications in the analysis - on the other hand, this releasing of constraints (theoretically allowing a much wider class of manipulation with simpler mechanisms) is exactly what makes dextrous manipulation with fingers such an attractive goal!

Thus it is immediately clear that a clear understanding of the nature of contact conditions (the geometry and physics of the constraints imposed between classes of fingertips and objects in contact) is a critical prerequisite for the development of motion and control algorithms for multifingered hands. Analysis of contact conditions and geometries has been the subject of considerable research in the community. A brief review of these efforts follows.

5.2.1 Contact conditions and the release of Constraints

From the above discussion it is clear that for multifingered grasping, a critical issue is the knowledge and modeling of the contact conditions present for a particular hand and held object. The existence of unconstrained degrees of freedom between the fingers and a held object allow rolling (relative rotational motion between the bodies) and sliding contacts (relative translational motion), and/or combinations of the two. This extra freedom in the contact conditions for fingers in general allows the possibility of more sophisticated manipulation than in the cooperating arm case, but at the cost of more complex planning and control requirements [114].

Significant early work concentrated on the kinematic constraints imposed at a contact, for different types of fingertip and object geometry [7]. Frictionless and frictional cases were explored. For example, a frictionless point contact (hard finger) model formally constrains only one direction of motion, where a soft finger contact (with friction) constrains at least four. Ultimately, complete tables have been set up detailing the kinematic constraints for different geometries [81].

The imposition of constraints by the existence of non-trivial contact conditions also complicates the static and dynamic analysis. In contrast to the cooperating arm case, fingers cannot impart forces and moments in arbitrary directions at the contact point. For example, a point contact model (hard finger) allows only forces to propagate through the contact points [15], where a soft finger constraint permits some moments to propagate. This of course again complicates the planning and control of multifingered grasps. However, since the static constraints imposed for a given finger/object contact are dual to the kinematic constraints, they can be detailed in a similar fashion.

At this time, the modeling of contact conditions and their constraints is a fairly well understood issue [81]. This is important since the effects of non-trivial contact conditions pervade almost all aspects of multifingered grasping research as we will see. In the following section, we review some of the areas of multifingered robot grasping that have occupied significant attention in the last decade or so.

5.3 Ongoing research issues

Recent research efforts in multifingered robot hands can be broken down into several themes, according to which sub-problem of multi-fingered manipulation is being addressed. In the following, we attempt to present an overview of the main research themes.

5.3.1 Grasp synthesis

The first natural question to investigate for multifingered hands involves how to configure the fingers of the hand when grasping an object. This is the problem of grasp synthesis, or grasp planning, and can be restated as 'at which points on the object should the fingers be placed?'. Notice that this is an issue that is 'natural' to humans, who grasp most objects instinctively. However, for robot hands (some of which have very different kinematic arrangements of the fingers than human hands) this is a non-trivial issue. Many researchers have concentrated on grasp synthesis [26, 30, 84, 103, 122] and planning [10, 25, 41, 43, 106, 140]. Much of this work has focused on matching the geometry of the hand to that of an object to be grasped.

Additional work has focused on grasp analysis [38, 64, 101, 110]. Various grasp quality measures have been proposed [21, 75, 121], in order to rate different possible grasp choices.

For example, in [75], Li *et al.* define three different grasp quality measures. Based on a definition of stability requiring the grasp geometry to allow the fingers to balance disturbance forces in all directions (under friction), a worst case grasp measure in [75] was based on the smallest singular value of the Grasp Matrix (which will be discussed in more detail in the next section). A second grasp measure was defined in [75] as the volume (in object space) of object forces and moments which were achievable with reasonable finger forces.

These two measures are functions purely of the geometry of the grasp. Finally, a third grasp measure in [75] was defined to incorporate the desired task into the description. For this measure, an alignment condition between an ellipsoid (representing the task) in object space and an ellipsoid derived from the grasp geometry (representing the ability of the grasp to manipulate in different directions) was evaluated. Grasps with closer alignments are to be preferred. For more details, see [75].

Some measures developed for use in other robotics scenarios have been adapted to the multifingered case. In the same way that manipulability and force ellipsoids, which give a geometric sense of the quality of a robot configuration, have been extended to the multiple armed case [23, 24], dynamic impact ellipsoids can be defined for multifingered grasps [129]. It is shown in [129] how these ellipsoids can effectively and intuitively distinguish between 'good' and 'bad' grasps from the point of view of impact.

One notion underlies much of the above work, the notion of grasp stability. Clearly it is usually desirable to choose a grasp that is 'stable', in some sense, in order to maintain the grasp of an object, possibly under external disturbances. Evaluation of grasps leads naturally to the issue of grasp stability [40, 49, 86, 88, 123], which can be expressed in several ways. A fundamental question in this regard is that of how many fingers are necessary in order to stably grasp a given object, and where these fingers should be placed. This is perhaps the area of multifingered hand research in which the most complete body of underlying theory has been developed. Some of the basic results are reviewed in the following.

5.3.2   Grasp stability

Key questions in this area include the issue of how many fingers or contacts are required to constrain a given object, under various contact conditions (frictionless point contact, etc.) Significant work in this area has established bounds on the number and type of contacts [80, 94, 108].

The definitions of Force Closure and Form Closure Grasps have emerged from these works in the last several years [6, 104, 107, 137]. At this time, the definitions of Form and Force Closure, and their interrelationship are the subject of strong debate. However, one definition that seems to be generally accepted [7] defines Form Closure (or complete constraint) as the ability of a grasp to prevent motions of the object, relying on only unilateral, frictionless contact constraints. Force Closure, on the other hand, is defined in [7] as the situation where motions of the object are constrained by suitably large contact forces of the grasp (usually considering friction).

As an example consider the Figure 5.3. The figure shows a three-fingered grasp of a planar circle, or disk. The grasp is not form closure in the sense above since a moment about the center of the circle can not be resisted by the fingers (with frictionless contact). However, the grasp is force closure under friction, since in this case the fingers can 'squeeze' suitably to invoke sufficient tangential frictional forces at the contact points to resist the moment at the disk center (and also all other planar disturbance forces and moments).

Using the above definition of form closure, Markenscoff *et al.* [80] show that form closure of any two-dimensional object with piecewise smooth boundary (except a circle, note the disk example above) can be achieved with four fingers. For three dimensions, it is shown in [80] that under very general conditions, form closure of any bounded object can be achieved with 7 fingers (provided the object does not have a rotational symmetry)

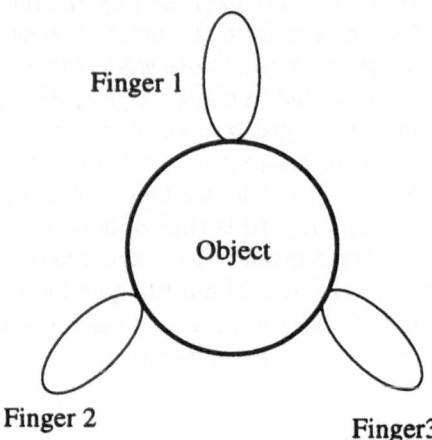

Figure 5.3: A Force Closure but not Form Closure Grasp.

These bounds seem a little excessive. However, when Coulomb friction is taken into account, it is shown in [80] that under the most relaxed assumptions three fingers are necessary and sufficient in two dimensions, and four fingers in three dimensions. This agrees with our intuition from the disk example above.

More recent significant work has considered the effects on form and force closure of second order (acceleration level) models [49, 107, 108]. This work has added increased understanding of the underlying physical effects of form and force closure, in particular focusing on conditions for the complete immobilization of an object, which can not be completely characterized by first order theories.

5.3.3 The importance of friction

From the above example, we see how helpful friction is in reducing the number of fingers theoretically necessary for grasping. In fact, this agrees strongly with our intuition. Humans perform dextrous grasping every day with as few as two fingers. This reduction in the number of required fingers over the above (worst case) bounds is largely due to our heavy reliance on friction at our fingertips.

In many robotics applications, this is not so easy to do, since fine control of frictional forces requires good sensing of effects such as slip [9], and such sensors are not readily available for robot hands at this time. Thus the above results are important primarily in establishing bounds on the

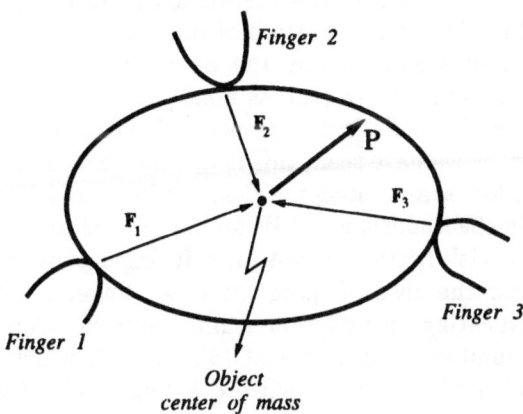

Figure 5.4: End-effector forces at contact points and object center of mass.

required number of fingers, and in guiding their positioning on the object. Notice however that friction is a practical ally in multifingered manipulation, though as we will see in the next section, this transfers the difficulty of modeling frictional constraints to the user.

5.3.4   Finger force distribution issues

In addition to the desire to constrain a held object when grasped, an important consideration is to plan and control the interactive coupling effects felt by the fingers through the object during manipulation. The desire to plan grasps that both constrain and/or manipulate a held object and also produce desired internal finger forces (squeezing) leads to the grasp force distribution problem. The problem, which is an extension of that for load distribution of cooperating arms, can be expressed as follows:

The total object inertial force can be expressed in terms of the end-effector forces as (see Figure 5.4).

$$\underline{P} = [W]\underline{F} \qquad (5.1)$$

where $\underline{P}$ is given by $[\underline{f}^T \underline{n}^T]^T$ and is the force and moment, respectively, experienced at the center of mass of the object (see Figure 5.4), and $\underline{F}$ is given by $[\underline{F}_1^T .....\underline{F}_L^T]^T$, the vector of forces (and moments) imparted to the object by the manipulators at the L contact points (note that T denotes transpose).

The matrix W is known in the literature as the *Grasp Matrix* or *Grip Matrix*. It is a function of the location of the contact points on the surface of the object. It thus incorporates the knowledge of the grasp geometry [81]. The Grasp matrix $[W]$ in (5.1) is nonsquare, of dimensions $6 \times 6L$ in general, if each of the L fingers imparted 3 forces and 3 moments to the object. However, as we have seen, this is not the case in general for fingers. If the number of forces and moments that can be transmitted through each contact is d, then the dimensions of W for the spatial case are $6 \times Ld$ ($d = 3$ for point contact with friction, $d = 4$ for soft finger contact)[1].

In general, for the case of more than two fingers, the Grasp matrix is nonsquare, indicating an underdetermined system. Consequently, there are an infinite number of solutions of (5.1) for $\underline{F}$, which corresponds to the infinite number of ways in which the L fingers can divide the motion task ('share the load') between themselves. The load distribution task is to choose the 'best' of these alternatives.

The general solution to (5.1) is given by

$$\underline{F} = [W^+]\underline{P} + [I - W^+W]\underline{\varepsilon} \tag{5.2}$$

where $[W^+]$ is a generalized inverse, or pseudoinverse, of $[W]$, I is the $Ld \times Ld$ identity matrix, and $\underline{\varepsilon}$ is an arbitrary vector whose values dictate which of the possible solutions of (5.1) for $\underline{F}$ is chosen. Equation (5.2) represents the basis for the great majority of approaches, both theoretical and empirical, to load distribution, and many algorithms to calculate $\underline{\varepsilon}$ for different possible pseudoinverses of $[W]$ have been suggested. There has been much work on this problem in the last few years [20, 68, 89, 99]. Most of the work concentrates on, for a given grasp configuration, solving for two components of finger forces: (a) a manipulating finger force component which constrains and moves the object as desired; and (b) an interactive finger force component which does not move the object, but generates internal forces on the object in an appropriate way.

The solution to the load distribution problem for multifingered hands is not as simple as directly finding a solution via (5.2), however. There are additional constraints, such as friction and the fact that the finger forces must be directed inwards towards the object (fingers can push but not pull). Thus the solution must satisfy (5.1) and the additional constraints (assuming static friction with friction coefficient μ):

- Pushing

$$f_{ni} = n_i' \cdot f_i \geq 0 \tag{5.3}$$

---

[1]In a more general case, the dimensions of W would be $6 \times (3a + 4b)$ where a is the number of point contacts with friction and b is the number of soft finger contacts

where n_i' is the normal to the plane of contact between finger i and the object.

- Friction

$$|f_{ti}| = \sqrt{f_i \cdot f_i - |f_{ni}|^2} \le \mu |f_{ni}| \tag{5.4}$$

where $F_i = [f_i^T n_i^T]^T$, $f_i = f_{ni} + f_{ti}$ and f_{ni} and f_{ti} are the normal and tangential (to the object surface) components of the applied finger force, respectively.

The force distribution problem has been solved including the friction constraints in various ways (see above references, and also [70, 138]). In general, at this time there are a variety of possible approaches to solving the finger force distribution problem, and this area is one of the better understood in multifingered grasp analysis.

5.3.5 Varying contacts: Rolling and sliding

Much of the above work has concentrated on analyzing candidate grasps singly (i.e. concentrating on one grasp in which the finger positions remain fixed to the same points on the held object during the analysis). However, there has also been much work on regrasping from one distinct grasp configuration to another [27, 81, 95]. For the case of regrasping by successive fingers discretely changing position on a grasped object, this is known as finger gaiting [7].

A significant body of work has also been built up in developing the theory of continuously evolving grasps, both for rolling [8, 28, 73, 139] and sliding [4, 9, 51, 61, 62, 69, 124, 126, 136] contacts. For the case of rolling contact, the fundamental work of Cai and Roth [15] and Montana [87] on the kinematics of contact has proved important in relating the evolution of contact positions on two bodies in contact to the velocity differences between the bodies. Montana's result, which is reviewed briefly below, has been the basis for much work in analyzing rolling contacts for multifingered hands.

5.3.6 Kinematics of rolling contact

In order to model and subsequently control rolling fingertip contacts of an object, it is desirable to keep track of several fundamental quantities: the object location, the fingertip contact locations, and the curves traced by the fingertips on the object. In [14] and [16], the authors derive relationships between velocities and higher order derivatives for planar and spatial curves in point contact. Montana [87] has derived a set of input-output equations which describe how the points of contact on the surfaces of the contacting bodies evolve in time in response to relative motion between the bodies, at the velocity level. Corresponding second-order relations have been obtained in [112]. The problem of determining the existence of an admissible path between two contact configurations and determining such a path, for rolling constraint has been studied in [73].

In this section, we briefly summarize the first-order contact kinematics derived by Montana. Montana's equations use the curvature, torsion and metric forms of the contacting surfaces (see [87] for more details) to relate the relative velocities between the two contacting bodies to the velocities of the contact points on each of the surfaces as follows.

The (instantaneous) relative motion between the bodies is defined as follows. Let v_x, v_y and v_z be the components of the translational velocity and ω_x, ω_y and ω_z be the components of the angular velocity of the finger with respect to the object at time t.

There are five degrees of freedom of the evolution of the contact points (one degree of freedom, normal to the plane of contact between the two bodies, is constrained by the contact), defined as follows. The quantities $\dot{u}_f$ and $\dot{u}_o$ are the (two-dimensional) velocities instantaneously tangential to the curves traced by the contact point on the finger and object, respectively. The angle of contact between the finger and object, $\psi(t)$ is measured about the normal to the plane of contact between the two bodies.

Let the curvature form, connection form and the metric of the finger surface and object surface finger/object contact point at time t, be $\mathcal{K}_f$, $\mathcal{T}_f$, $\mathcal{M}_f$ and $\mathcal{K}_o$, $\mathcal{T}_o$, $\mathcal{M}_o$ respectively. Also, let

$$R_\psi = \begin{bmatrix} \cos\psi & -\sin\psi \\ -\sin\psi & -\cos\psi \end{bmatrix} \quad ; \quad \tilde{\mathcal{K}}_o = R_\psi \mathcal{K}_o R_\psi$$

$(\mathcal{K}_f + \tilde{\mathcal{K}}_o)$ is called the relative curvature form.

At a point of contact, if the relative curvature form is invertible, then the point of contact and angle of contact evolve according to the following equations

$$\dot{u}_f = \mathcal{M}_f^{-1}(\mathcal{K}_f + \tilde{\mathcal{K}}_o)^{-1} \left(\begin{bmatrix} -\omega_y \\ \omega_x \end{bmatrix} - \tilde{\mathcal{K}}_o \begin{bmatrix} v_x \\ v_y \end{bmatrix} \right) \tag{5.5}$$

$$\dot{u}_o = \mathcal{M}_o^{-1} R_\psi (\mathcal{K}_f + \tilde{\mathcal{K}}_o)^{-1} \left(\begin{bmatrix} -\omega_y \\ \omega_x \end{bmatrix} + \mathcal{K}_f \begin{bmatrix} v_x \\ v_y \end{bmatrix} \right) \tag{5.6}$$

$$\dot{\psi} = \omega_z + \mathcal{T}_f \mathcal{M}_f \dot{u}_f + \mathcal{T}_o \mathcal{M}_o \dot{u}_o \tag{5.7}$$

$$v_z = 0 \tag{5.8}$$

In particular, if the bodies maintain rolling contact with each other, this implies that the relative translational velocities are zero, i.e.

$$\left[\begin{array}{c} v_x \\ v_y \end{array} \right] = 0 \tag{5.9}$$

Additionally, if the bodies are not allowed to spin with respect to each other (*pure rolling* motion), then

$$\omega_z = 0 \tag{5.10}$$

Substituting conditions (5.9) and (5.10) in the kinematic contact equations (5.5–7), we obtain the first order equations for pure rolling contact as

$$\left[\begin{array}{c} \dot{\mathbf{u}}_f \\ \dot{\mathbf{u}}_o \\ \dot{\psi} \end{array} \right] = \left[\begin{array}{c} \mathcal{M}_f^{-1} \\ \mathcal{M}_o^{-1} \\ \mathcal{T}_f + \mathcal{T}_o R_\psi \end{array} \right] \left(\mathcal{K}_f + \tilde{\mathcal{K}}_o \right)^{-1} \left[\begin{array}{c} -\omega_y \\ \omega_x \end{array} \right] \tag{5.11}$$

Much of the work in evolving rolling grasps has built on this framework, combining the above model with the conventional dynamic models of the object and fingers. With rolling contact, an important point to note is that the situation is complicated by the fact that the motion planning problem is inherently nonholonomic [91]. Various methods (see above, and also [28, 32, 73, 113, 139]) have been proposed to address this issue. At this time, perhaps largely due to the computational complexity of the models involved, most of this work has been performed in simulation, rather than on actual hardware.

In the case of sliding between the fingertips and object, a distinction is drawn between the frictionless and non-zero friction cases. In the case of friction, enough tangential finger force must be applied to begin sliding. There is the notion of contact formations, which describe functionally the state of a grasp, by annotating which geometrical features (edges, faces, etc) of the object are in contact with those of which fingers. Grasps which are equivalent under sliding are identified by the same contact formation. There has been significant work in understanding when it is possible to move from one to another distinct contact formations by fingertip sliding. The investigation of manipulation by sliding is currently an active area of research [4, 9, 51, 61, 62, 69, 124, 126, 136].

5.3.7 Grasp compliance and control

Given a grasp analysis/plan, there has been extensive work in the area of grasp control [3, 45, 52, 74, 92, 98, 118] and optimization [12]. Real-time control of robot hands is made difficult by the complexity of the dynamic models, and the difficulty of extracting good sensory data in real-time from typical hands. A good approach which has been used by a number of

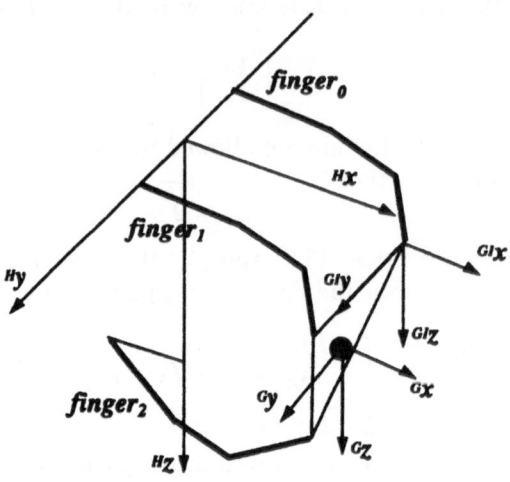

Figure 5.5: Hand grasp frame.

researchers is impedance control [33], or stiffness control [13, 118]. For example, in [33] the following layered impedance strategy is successfully employed for the JSC Salisbury hand. Impedance loops control tendon tension, joint moments, Cartesian finger forces, and grasped object forces and moments. At the lowest level tendon tension levels are obtained by driving each motor in velocity mode. The tension levels are read at strain gauges located directly behind the finger assemblies ensuring accurate tension control at a distance from the driving motors.

For thread mating experiments in [33] the Salisbury hand is configured in object grasp control mode, where the three fingers come together to grasp a male threaded fastener. In this configuration, the hand's 9 degrees of freedom combine together to actuate 6 degrees of freedom rigid body object control at the grasp center which is located at the centroid of the triangle defined by the finger tips (Figure 5.5). The three remaining degrees of freedom are used to maintain a positive force between each pair of fingers to prevent slip. Object force control is maintained by commanding the fingers in concert to yield forces and moments at the grasp center. The actual object force command, $\underline{Q}$, is generated from position errors in the grasp frame to yield a stiffness control of the form:

$$Q = [K]\underline{e}, \tag{5.12}$$

where the total force vector $\underline{Q}$ is given by:

$$\underline{Q} = \begin{bmatrix} f_x & f_y & f_z & \tau_{roll} & \tau_{pitch} & \tau_{yaw} \\ f_{01} & f_{02} & f_{12} \end{bmatrix}^T. \tag{5.13}$$

$[K]$ is a 9 x 9 diagonal matrix and its elements can be set to obtain an arbitrary stiffness in each axis, and

$$\underline{e} = \begin{bmatrix} e_x & e_y & e_z & e_{roll} & e_{pitch} & e_{yaw} \\ e_{x01} & e_{x02} & e_{x12} \end{bmatrix}^T, \tag{5.14}$$

The rigid body elements of the object force, $\underline{Q}$ and the object position error, $\underline{e}$ have their standard meanings (fixed angles). The force element, f_{01} refers to the force between fingers 0 and 1 along a vector between the tips, and f_{02} and f_{12} follow the same convention. The position error element e_{x01} is the error in the distance between finger tips 0 and 1.

The above approach largely neglects much of the compliance physically present in the hand itself. There is a general need for more complete compliance models. The issue of grasp compliance (that of determining the overall effective compliance of a hand and object) has addressed significant attention recently [31, 50, 125].

5.4 Further research issues

Much of the work in analysis of multifingered robot grasping draws, at least intuitively, on features of human grasping. There has been work in the analysis of human hands and fingers [5, 42, 44, 60] and application to both robotics and prosthetics [37].

One feature of human grasping is that in many grasps, not only the fingertips are used (as has been the case in most robotic hand analysis and experiments). This type of grasp is typically denoted a precision grasp. Recent work has begun to address the issue of full finger and power grasps for robot hands [50, 54, 66, 85, 96, 141, 142]. In power grasps, grasps are made as in Figure 5.6, with contact between the object and the intermediate finger joints, as well as the fingertips and possibly the palm. This type of grasp is inherently more stable than fingertip grasps, however analysis is more difficult due to the extra constraints (and inherent loss of degrees of freedom) from the additional contacts.

In contrast, there is recent interest in the issue of manipulation without grasping, or nonprehensile manipulation [144]. This offers the possibility of using simpler mechanisms to achieve the necessary results with the minimum hardware (minimalist robotics). There is a strong relationship here

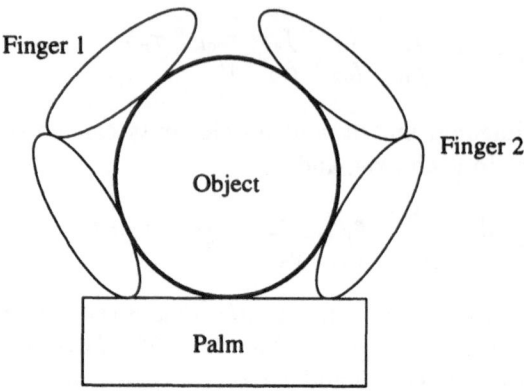

Figure 5.6: Full fingered power grasp.

to the problem of parts sorting [1, 59, 76]. In this case, the difficulty is shifted from the device design to the skill and creativity of the planner.

This point of view seems set to generate much interest, since most of the robot hands built thus far still suffer from being underutilized in the sense of having more physical capability (in theory) than is utilized with current sensing and planning methods. Other biologically-inspired attempts to utilize simpler hand designs include the analysis of simple but effective multifingered hand designs in nature [130] and recent application of genetic programming to grasp synthesis [39].

5.5 Current limitations

One important limitation of current multifingered grasping strategies is that most demonstrations have involved manipulating a held object independent of the rest of the world. Very few multifingered hands have been used effectively as end effectors [83], contacting and recontacting the environment. However, the fundamental nature of many interesting manipulation tasks is that they involve contact (basically impact) with a partially modeled environment. There is a need for more research in this area, although there has been some successful early work [53, 130].

In addition, the vast majority of work in the multifingered grasping area to date has been theoretical in nature. Many of the algorithms developed have not yet been successfully applied to hardware. There has however been a steady increase in experimental work in the last few years [11, 34, 56].

There are also steady improvements in the areas of hand design, sensing, programming and control environments for robot hands. However, the state of the art multifingered robot hand is still, in general, fairly difficult to program and control, suffers from an insufficiently rich choice of sensors, and is fairly unreliable.

5.6 Conclusions

This chapter has concentrated mostly on theoretical and algorithmic developments. However, there is ongoing progress in hardware, particularly for sensing and control. In particular, there is a critical need for better sensory systems (both contact and non-contact) in robot hands, in order to sense and learn about the environment and grasp objects. Most of the work mentioned above assumes significant knowledge of the object to be manipulated and the environment. However, there is ongoing work in grasping in uncertain environments, [58, 90, 117], using learning and knowledge-based systems. Recent work has concentrated on various aspects of multifingered manipulation incorporating tactile feedback [17, 18, 19, 48, 57, 78, 79, 115].

In summary, there has been much progress in the area of multifingered robot hands and dextrous robot grasping in the past few years. Most of the progress has been in the area of analysis of grasps and their evolution, though there is now a strong body of experimental work to support much of the theory. However, the results are still basically constrained to research laboratories, and significant applications of multifingered robot hands are lacking. More work is needed in the reliability of hand designs, sensing technologies (especially), and programming and control environments, in addition to further progress in the theoretical area. The next few years promise to be an exciting time in the area. For further information and sources, the reader is encouraged to consult [7, 29, 42, 63, 91, 100].

Acknowledgements

This work was supported in part by the National Science Foundation under grants CMS-9532081 and IRI-9526363, by the Office of Naval Research under contract N00014-06-C-0320, and by NASA contract NAG-9-740.

References

[1] S. Akella and M.T. Mason. Parts Orienting by Push-Aligning. In *Proceedings of the 1995 IEEE International Conference on Robotics*

and Automation, pages 414–420, 1995.

[2] M.S. Ali, K.J. Kyriakopoulos, and H.E. Stephanou. The Kinematics of the Anthrobot-2 Dextrous Hand. In *Proceedings 1993 IEEE International Conference on Robotics and Automation*, pages 705–710, Atlanta, GA, 1993.

[3] P.K. Allen, M. Paul, and K.S. Robots. A System for Programming and Controlling a Multisensor Robotic Hand. *IEEE Transactions on Robotics and Automation*, 20(6):1450–1455, 1990.

[4] B. Barkat, G. Bessonet, and J.P. Lallemand. Hyper-Static Grasping Optimization with Finger Deformability and Sliding Constraints. In *Proceedings 1994 IEEE International Conference on Robotics and Automation*, pages 1923–1930, San Diego, CA, 1994.

[5] G.A. Bekey, H. Liu, R. Tomovic, and W.J. Karplus. Knowledge-Based Control of Grasping in Robot Hands using Heuristics from Human Motor Skills. In *Proceedings 1993 IEEE International Conference on Robotics and Automation*, pages 709–722, Atlanta, GA, 1993.

[6] A. Bicchi. On the Closure Properties of Robotic Grasping. *International Journal of Robotics Research*, 14(4):x–x, 1995.

[7] A. Bicchi. Hands for Dexterous Manipulation and Powerful Grasping: A Difficult Road Towards Simplicity. In *1988 IEEE International Conference on Robotics and Automation: Workshop on Minimalism in Robot Manipulation*, pages 1–13, Minneapolis, MN, 1996.

[8] A. Bicchi and R. Sorrentino. Dexterous Manipulation through Rolling. In *Proceedings 1995 IEEE International Conference on Robotics and Automation*, pages 452–457, Nagoya, Japan, 1995.

[9] D.L. Brock. Enhancing the Dexterity of a Robot Hand using Controlled Slip. In *Proceedings 1988 IEEE International Conference on Robotics and Automation*, pages 249–251, Philadelphia, PA, 1988.

[10] R.C. Brost. Automatic grasp planning in the presence of uncertainty. *International Journal of Robotics Research*, 7(1):3–17, 1988.

[11] M. Buss and H. Hashimoto. Dextrous Robot Hand Experiments. In *Proceedings 1995 IEEE International Conference on Robotics and Automation*, pages 1680–1686, Nagoya, Japan, 1995.

[12] M. Buss, H. Hashimoto, and J.B. Moore. Grasping Force Optimization for Multifingered Robot Hands. In *Proceedings 1995 IEEE International Conference on Robotics and Automation*, pages 1034–1039, Nagoya, Japan, 1995.

[13] M. Buss and K.P. Kleinmann. Multi-Fingered Grasping Experiments Using Real-Time Grasping Force Optimization. In *Proceedings 1996 IEEE International Conference on Robotics and Automation*, pages 1807–1812, Minneapolis, MN, 1996.

[14] C. Cai and B. Roth. On the planar motion of rigid bodies with point contact. *Mechanism and Machine Theory*, 21(6):453–466, 1986.

[15] C. Cai and B. Roth. On the Spatial Motion of a Rigid Body with Point Contact. In *Proceedings 1987 IEEE International Conference on Robotics and Automation*, pages 686–695, Raleigh, NC, 1987.

[16] C. Cai and B. Roth. On the spatial motion of a rigid body with point contact. In *Proceedings of the 1987 IEEE International Conference on Robotics and Automation*, pages 686–695, 1987.

[17] D.G. Caldwell, S. Lawther, and A. Wardle. Tactile Perception and its Application to the Design of Multi-modal Cutaneous Feedback Systems. In *Proceedings 1996 IEEE International Conference on Robotics and Automation*, pages 3215–3221, Minneapolis, MN, 1996.

[18] M. Charlebois, K. Gupta, and S. Payandeh. Curvature Based Shape Estimation Using Tactile Sensing. In *Proceedings 1996 IEEE International Conference on Robotics and Automation*, pages 3502–3507, Minneapolis, MN, 1996.

[19] N. Chen, R. Rink, and H. Zhang. Local Object Shape from Tactile Sensing. In *Proceedings 1996 IEEE International Conference on Robotics and Automation*, pages 3496–3501, Minneapolis, MN, 1996.

[20] Y.C. Chen, I.D. Walker, and J.B. Cheatham. A New Approach to Force Distribution and Planning for Multifingered Grasps of Solid Objects. In *Proceeedings 1991 IEEE Conference on Robotics and Automation*, pages 890–897, 1991.

[21] Y.C. Chen, I.D. Walker, and J.B. Cheatham. Visualization of Force-Closure Grasps for Objects Through Contact Force Decomposition. *International Journal of Robotics Research*, 14(1):37–75, 1995.

[22] F.T. Cheng and D.E. Orin. Efficient Algorithm for Optimal Force Distribution - The Compact-Dual LP Method. *IEEE Transactions on Robotics and Automation*, 6(2):178–187, 1990.

[23] P. Chiacchio, S. Chiaverini, L. Sciavicco, and B. Siciliano. Global Task Space Manipulability Ellipsoids for Multiple Armed Systems. *IEEE Transactions on Robotics and Automation*, RA-7(5):678–685, 1991.

[24] P. Chiacchio, S. Chiaverini, L. Sciavicco, and B. Siciliano. Task Space Dynamic Analysis of Multiarm System Configurations. *International Journal of Robotics Research*, 10(6):708–715, 1991.

[25] N.Y. Chong, D.H. Choi, and I.H. Suh. Dextrous Manipulation Planning of Multifingered Hands with Soft Finger Contact Model. In *Proceedings 1994 IEEE International Conference on Robotics and Automation*, pages 3389–3396, San Diego, CA, 1994.

[26] J.A. Coelho and R.A. Grupen. Online Grasp Synthesis. In *Proceedings 1996 IEEE International Conference on Robotics and Automation*, pages 2137–2142, Minneapolis, MN, 1996.

[27] A.A. Cole, P. Hsu, and S. Sastry. Dynamic Regrasping by Coordinated Control of Sliding for a Multifingered Hand. In *Proceedings 1989 IEEE International Conference on Robotics and Automation*, pages 781–786, Scottsdale, AZ, 1989.

[28] A.C. Cole, J.E. Hauser, and S. Sastry. Kinematics and Control of Multifingered Hands with Rolling Contact. *IEEE Transactions on Automatic Control*, AC-34(34):398–404, 1989.

[29] M.R. Cutkosky. *Robotic Grasping and Fine Manipulation*. Kluwer, Boston, 1985.

[30] M.R. Cutkosky. On grasp choice, grasp models, and the design of hands for manufacturing tasks. *IEEE Transactions on Robotics and Automation*, RA-5(3):269–279, 1989.

[31] M.R. Cutkosky and I. Kao. Computing and Controlling the Compliance of a Robotic Hand. *IEEE Transactions on Robotics and Automation*, 5(2):151–165, 1989.

[32] A.S. Deo and I.D. Walker. Dynamics and Control Methods for Cooperating Manipulators with Rolling Contacts. In *Proceedings IEEE International Conference on Robotics and Automation*, pages 1509–1516, Nagoya, Japan, 1995.

[33] M.A. Diftler and I.D. Walker. Experiments in Mating Threaded Parts Using a Dexterous Robotic Hand. In *Proceedings of the 1996 IEEE-SMC Symposium on Robotics and Cybernetics, CESA '96*, pages 459–464, 1996.

[34] R.A. Diftler and I.D. Walker. Determining Alignment between Threaded Parts Using Force and Position Data from a Robot Hand. In *Proceedings 1997 IEEE International Conference on Robotics and Automation*, pages 1503-1510, Albuquerque, NM, 1997.

[35] German Aerospace Research Establishment (DLR). *Status Report*. Institute of Robotics and System Dynamics, 1997.

[36] M. Ebner and R.S. Wallace. A Direct-Drive Hand: Design, Modeling, and Control. In *Proceedings 1995 IEEE International Conference on Robotics and Automation*, pages 1668–1674, Nagoya, Japan, 1995.

[37] K.A. Farry, I.D. Walker, and R.G. Baraniuk. Myoelectirc teleoperation of a Complex Robotic Hand. *IEEE Transactions on Robotics and Automation*, 12(5):775–788, 1996.

[38] R. Fearing. Simplified Grasping and Manipulation with Dextrous Robot Hands. In *Proceedings 1986 IEEE International Conference on Robotics and Automation*, pages 188–195, San Francisco, CA, 1986.

[39] J.J. Fernandez and I.D. Walker. Biologically Inspired Control for Semi-Autonomous Robotic Grasping. In *Workshop on Evolutionary Robotics, 1997 International Conference on Genetic Algorithms*, page 1, East Lansing, MI, 1997.

[40] Y. Funahashi, T. Yamada, M. Tate, and Y. Suzuki. Grasp Stability Analysis Considering the Curvatures at Contact Points. In *Proceedings 1996 IEEE International Conference on Robotics and Automation*, pages 3040–3046, Minneapolis, MN, 1996.

[41] R.A. Grupen. Planning Grasp Strategies for Multifingered Robot Hands. In *Proceedings 1991 IEEE International Conference on Robotics and Automation*, pages 646–651, Sacramento, CA, 1991.

[42] R.A. Grupen, T.C. Henderson, and I.D. McCammon. A survey of general-purpose manipulation. *International Journal of Robotics Research*, 8(1):38–62, 1989.

[43] G. Guo, W.A. Gruver, and K. Jin. Grasp Planning for Multi-fingered Robot Hands. In *Proceedings 1992 IEEE International Conference on Robotics and Automation*, pages 2284–2289, Nice, France, 1992.

[44] H-Y. Han, A. Shimada, and S. Kawamura. Analysis of Friction on Human Fingers and Design of Artificial Fingers. In *Proceedings 1996 IEEE International Conference on Robotics and Automation*, pages 3061–3066, Minneapolis, MN, 1996.

[45] H. Hashimoto, H. Ogawa, T. Umeda, M. Obama, and K. Tatsuno. An Unilateral Master-Slave Hand System with a Force-Controlled Slave Hand. In *Proceedings 1995 IEEE International Conference on Robotics and Automation*, pages 1668–1674, Nagoya, Japan, 1995.

[46] S. Hayati. Hybrid position/force control of multi-arm cooperating robots. In *Proceedings 1986 IEEE Conference on Robotics and Automation*, pages 82–89, 1986.

[47] C. Hess, L.C.H. Li, K.A. Farry, and I.D. Walker. Application of Dexterous Space Robotics Technology to Myoelectric Prostheses. In *Proceedings Technology 2003 Conference, NASA Conf. Publication 3249*, pages 255–268, Anaheim, CA, 1993.

[48] E.G.M. Holweg, H. Hoeve, W. Jongkind, L. Marconi, C. Melchiorri, and C. Bonivento. Slip Detection by Tactile Sensors: Algorithms and Experimental Results. In *Proceedings 1996 IEEE International Conference on Robotics and Automation*, pages 3234–3239, Minneapolis, MN, 1996.

[49] W.S. Howard and V. Kumar. Stability of Planar Grasps. In *Proceedings 1994 IEEE International Conference on Robotics and Automation*, pages 2822–2827, San Diego, CA, 1994.

[50] W.S. Howard and V. Kumar. Modeling and Analysis of the Compliance and Stability of Enveloping Grasps. In *Proceedings 1995 IEEE International Conference on Robotics and Automation*, pages 1367–1372, Nagoya, Japan, 1995.

[51] R. Howe, I. Kao, and M.R. Cutkosky. The Sliding of Robot Fingers under Combined Torsion and Shear Loading. In *Proceedings 1988 IEEE International Conference on Robotics and Automation*, pages 103–105, Philadelphia, PA, 1988.

[52] P. Hsu, Z. Li, and S. Sastry. On Grasping and Coordinated Manipulation by a Multifingered Robot Hand. In *Proceedings 1988 IEEE International Conference on Robotics and Automation*, pages 384–389, Philadelphia, PA, 1988.

[53] W.H. Huang, E.P. Krotkov, and M.T. Mason. Impulsive Manipulation. In *Proceedings of the 1995 IEEE International Conference on Robotics and Automation*, pages 120–125, 1995.

[54] O.M. Ismaeil and R.E. Ellis. Grasping Using the Whole Finger. In *Proceedings 1994 IEEE International Conference on Robotics and Automation*, pages 3111–3116, San Diego, CA, 1994.

[55] S.C. Jacobsen, E.K. Iversen, D.F. Knutti, R.T. Johnson, K.B. Biggers. Design of the Utah/MIT Dextrous Hand. In *IEEE Conference on Robotics and Automation*, pages 1520–1532, San Francisco, CA, 1986.

[56] B.M. Jau. Dexterous Manipulation with Four-Fingered Hand System. In *Proceedings 1995 IEEE International Conference on Robotics and Automation*, pages 338–343, Nagoya, Japan, 1995.

[57] D. Johnston, P. Zhang, J. Hollerbach, and S. Jacobsen. A Full Tactile Sensing Suite for Dextrous Robot Hands and Use in Contact Force Control. In *Proceedings 1996 IEEE International Conference on Robotics and Automation*, pages 3222–3227, Minneapolis, MN, 1996.

[58] I. Kamon, T. Flash, and S. Edelman. Learning to Grasp Using Visual Information. In *Proceedings 1996 IEEE International Conference on Robotics and Automation*, pages 2470–2476, Minneapolis, MN, 1996.

[59] D. Kang and K. Goldberg. Sorting parts by Random Grasping. *IEEE Transactions on Robotics and Automation*, 11(1):146–152, 1995.

[60] I. Kao. Stiffness Control and Calibration of Robotic and Human Hands and Fingers. In *Proceedings 1994 IEEE International Conference on Robotics and Automation*, pages 399–406, San Diego, CA, 1994.

[61] I. Kao and M.R. Cutkosky. Quasistatic Manipulation with Compliance and Sliding. *International Journal of Robotics Research*, 11(1):20–40, 1992.

[62] I. Kao and M.R. Cutkosky. Comparison of Theoretical and Experimental Force/Motion Trajectories for Dextrous Manipulation with Sliding. *International Journal of Robotics Research*, 12(6):529–534, 1993.

[63] I. Kato and K. Sadamoto. *Mechanical Hands Illustrated*. Hemisphere publishers, Springer-Verlag, 1987.

[64] J. Kerr and B. Roth. Analysis of multifingered hands. *International Journal of Robotics Research*, 4(4):3–17, 1986.

[65] O. Khatib. Augmented and Reduced Effective Inertia in Robotic Systems. In *Proceedings 1988 American Control Conference*, pages 2140–2147, 1988.

[66] K.P. Kleinmann, J-O. Hennig, C. Ruhm, and H. Tolle. Object Manipulation by a Multifingered Gripper: On the Transition from Precision to Power Grasp. In *Proceedings 1996 IEEE International Conference on Robotics and Automation*, pages 2761–2766, Minneapolis, MN, 1996.

[67] K. Kreutz and A. Lokshin. Load Balancing and Closed Chain Multiple Arm Control. In *Proceedings 1988 American Control Conference*, pages 2148–2155, 1988.

[68] V. Kumar and K.J. Waldron. Suboptimal Algorithms for Force Distribution in Multifingered Grippers. *IEEE Transactions on Robotics and Automation*, 5(4):491–498, 1989.

[69] V. Kumar, X. Yun, E. Paljug, and N. Sarkar. Control of Contact Conditions for Manipulation with Multiple Robotic Systems. In *Proceedings 1991 IEEE International Conference on Robotics and Automation*, pages 170–175, Sacramento, CA, 1991.

[70] V.M. Kvrgic. Computing the Sub-optimal Grasping Forces for Manipulation of a Rough Object by Multifingered Robot Hand. In *Proceedings 1996 IEEE International Conference on Robotics and Automation*, pages 1801–1806, Minneapolis, MN, 1996.

[71] P. Larochelle and J.M. McCarthy. Performance Evaluation of Cooperating Robot Movements Using Maximum Load Under Time-Optimal Control. In *Proceedings 1992 IEEE International Conference on Robotics and Automation*, pages 2612–2617, Nice, France, 1992.

[72] S. Lee and J.M. Lee. Task-oriented dual-arm manipulability and its application to configuration optimization. In *Proceedings IEEE Conference on Decision and Control*, pages 2253–2260, 1988.

[73] Z. Li and J. Canny. Motion of two rigid bodies with rolling constraint. *IEEE Transactions on Robotics and Automation*, RA-6(1):62–72, 1990.

[74] Z. Li, P. Hsu, and S. Sastry. Grasping and coordinated manipulation by a multifingered robot hand. *International Journal of Robotics Research*, 8(4):33–50, 1989.

[75] Z. Li and S. Sastry. Task-oriented Optimal Grasping by Multifingered Robotic Hands. *IEEE Journal of Robotics and Automation*, RA-4(1):32–44, February 1987.

[76] K.M. Lynch and M.T. Mason. Controllability of Pushing. In *Proceedings of the 1995 IEEE International Conference on Robotics and Automation*, pages 112–119, 1995.

[77] K. Machida, Y. Toda, Y. Murase, and S. Komada. Precise Space Telerobotic System using 3-Finger Multisensory Hand. In *Proceedings 1995 IEEE International Conference on Robotics and Automation*, pages 32–38, Nagoya, Japan, 1995.

[78] H. Maekawa, K. Kanie, and K. Komoriya. Tactile Sensor Based Manipulation of an Unknown Object by a Multifingered nad with Rolling Contact. In *Proceedings 1995 IEEE International Conference on Robotics and Automation*, pages 743–750, Nagoya, Japan, 1995.

[79] H. Maekawa, K. Tanie, and K. Komoriya. Dynamic Grasping Force Control Using Tactile Feedback for Grasp of Multifingered Hand. In *Proceedings 1996 IEEE International Conference on Robotics and Automation*, pages 2462–2469, Minneapolis, MN, 1996.

[80] X. Markenscoff, L. Ni, and C.H. Papadimitriou. The Geometry of Grasping. *International Journal of Robotics Research*, 9(1):61–74, 1990.

[81] M.T. Mason and J.K. Salisbury. *Robot Hands and the Mechanics of Manipulation*. MIT Press, 1985.

[82] C. Melchiorri and G. Vassura. Mechanical and Control Features of the University of Bologna Hand Version 2. In *Proceedings 1995*

IEEE/RSJ International Conference on Intelligent Robotics and Systems (IROS92), pages 187–193, 1992.

[83] P. Michelman and P. Allen. Forming Complex Dextrous Manipulations from Task Primitives. In *Proceedings 1994 IEEE International Conference on Robotics and Automation*, pages 3383–3388, San Diego, CA, 1994.

[84] B. Mirtich and J. Canny. Easily Computable Optimum Grasps in 2-D and 3-D. In *Proceedings 1994 IEEE International Conference on Robotics and Automation*, pages 739–747, San Diego, CA, 1994.

[85] K. Mirza and D.E. Orin. General Formulation for Force Distribution in Power Grasp. In *Proceedings 1994 IEEE International Conference on Robotics and Automation*, pages 880–887, San Diego, CA, 1994.

[86] B. Mishra and N. Silver. Some Discussion of Static Grasping and Its Stability. *IEEE Transactions on Systems, Man, and Cybernetics*, 19(4):783–796, 1989.

[87] D.J. Montana. The Kinematics of Contact and Grasp. *International Journal of Robotics Research*, 7(3):17–32, 1988.

[88] D.J. Montana. Contact Stability for Two-Fingered Grasps. *IEEE Transactions on Robotics and Automation*, 8(4):421–430, 1992.

[89] S.B. Moon and S. Ahmad. Sub-Time-Optimal Trajectory Plannings for Cooperative Multi-Manipulator Systems Using the Load Distribution Scheme. In *Proceedings 1993 IEEE International Conference on Robotics and Automation (Vol. 1)*, pages 1037–1042, Atlanta, GA, 1993.

[90] M.A. Moussa and M.S. Kamel. An Experimental Approach to Robotic Grasping Using Reinforcement Learning and Generic Grasping Functions. In *Proceedings 1996 IEEE International Conference on Robotics and Automation*, pages 2767–2773, Minneapolis, MN, 1996.

[91] R.M. Murray, Z. Li, and S.S. Sastry. *A Mathematical Introduction to Robotic Manipulation*. CRC Press, 1994.

[92] K. Nagai and T. Yoshikawa. Dynamic Manipulation/Grasping Control of Multifingered Robot Hands. In *Proceedings 1993 IEEE International Conference on Robotics and Automation*, pages 1027–1032, Atlanta, GA, 1993.

[93] Y. Nakamura, K. Nagai, and T. Yoshikawa. Mechanics of Coordinative Manipulation by Multiple Robotic Mechanisms. In *Proceedings 1988 IEEE Conference on Robotics and Automation*, pages 991–998, 1988.

[94] V. Nguyen. Constructing force-closure grasps. *International Journal of Robotics Research*, 8(1):26–37, 1989.

[95] T. Omata and M.A. Farooqi. Regrasps by a Multifingered Hand based on Primitives. In *Proceedings 1996 IEEE International Conference on Robotics and Automation*, pages 2774–2780, Minneapolis, MN, 1996.

[96] T. Omata and K. Nagata. Rigid Body Analysis of the Indeterminate Grasp Force in Power Grasps. In *Proceedings 1996 IEEE International Conference on Robotics and Automation*, pages 1787–1794, Minneapolis, MN, 1996.

[97] D.E. Orin and S.Y. Oh. Control of Force Distribution in Robotic Mechanisms Containing Closed Kinematic Chains. *ASME Journal of Dynamic Systems, Measurement, and Control*, 102:134–141, 1981.

[98] P.D. Panagiotopoulos and A.M. Al-Fahed. Robot Hand Grasping and Related Problems: Optimal Control and Identification. *International Journal of Robotics Research*, 13(2):127–136, 1994.

[99] Y.C. Park and G.C. Starr. Finger force computation for manipulation of an object by a multifingered robot hand. In *Proceedings 1989 IEEE Conference on Robotics and Automation*, pages 930–935, 1989.

[100] M.A. Peshkin. *Robotic Manipulation Strategies*. Prentice-Hall, 1990.

[101] F. Pfeiffer and K. Richter. Optimal Path Planning Including Forces at the Gripper. *Journal of Intelligent and Robotic Systems*, 3:251–258, 1990.

[102] N.S. Pollard. Planning Grasps for a Robot Hand in the Presence of Obstacles. In *Proceedings 1993 IEEE International Conference on Robotics and Automation (Vol. 3)*, pages 723–728, Atlanta, GA, 1993.

[103] N.S. Pollard. Synthesizing Grasps from Generalized Prototypes. In *Proceedings 1996 IEEE International Conference on Robotics and Automation*, pages 2124–2130, Minneapolis, MN, 1996.

[104] J. Ponce, D. Stam, and B. Faverjon. On Computing Two-finger Force-Closure Grasps of Curved 2-D Objects. *International Journal of Robotics Research*, 12(3):263–273, 1993.

[105] K. Rao, G. Medioni, H. Liu, and G.A. Bekey. Robot Hand-Eye Coordination: Shape Description and Grasping. In *Proceedings of the 1988 IEEE International Conference on Robotics and Automation*, pages 407–411, 1988.

[106] D. Reznik and V. Lumelsky. Multi-Finger Hugging: A Robust Approach to Sensor-Based Grasp Planning. In *Proceedings 1994 IEEE International Conference on Robotics and Automation*, pages 754–759, San Diego, CA, 1994.

[107] E. Rimon and J. Burdick. On Force and Form Closure for Multiple Finger Grasps. In *Proceedings 1996 IEEE International Conference on Robotics and Automation*, pages 1795–1800, Minneapolis, MN, 1996.

[108] E. Rimon and J.W. Burdick. New Bounds on the Number of Frictionless Fingers Required to Immobilize 2D Objects. In *Proceedings 1995 IEEE International Conference on Robotics and Automation*, pages 751–757, Nagoya, Japan, 1995.

[109] R.N. Rohling and J.M. Hollerbach. Optimized Fingertip Mapping for Teleoperation of Dextrous Robot Hands. In *Proceedings 1993 IEEE International Conference on Robotics and Automation (Vol. 3)*, pages 769–775, Atlanta, GA, 1993.

[110] L. Romdhane and J. Duffy. Kinestatic Analysis of Multifingered Hands. *International Journal of Robotics Research*, 9(6):3–18, 1990.

[111] J.K. Salisbury and J.J. Craig. Articulated Hands: Force Control and Kinematic Issues. *International Journal of Robotics Research*, 1:4–17, 1982.

[112] N. Sarkar, X. Yun, and V. Kumar. Dynamic control of 3-D rolling contacts in two-arm manipulation. In *Proceedings of the 1993 IEEE International Conference on Robotics and Automation*, pages 978–983, 1993.

[113] N. Sarkar, X. Yun, and V. Kumar. Dynamic Control of 3-D Rolling Contacts in Two-Arm Manipulation. In *Proceedings 1993 IEEE International Conference on Robotics and Automation*, pages 978–983, Atlanta, GA, 1993.

[114] P.R. Sinha and J.M. Abel. A Contact Stress Model for Multifingered Grasps of Rough Objects. *IEEE Transactions on Robotics and Automation*, 8(1):2–22, 1992.

[115] J.S. Son and R.D. Howe. Tactile Sensing and Stiffness Control with Multifingered Hands. In *Proceedings 1996 IEEE International Conference on Robotics and Automation*, pages 3228–3233, Minneapolis, MN, 1996.

[116] T.H. Speeter. Control of the UTAH/MIT Dextrous Hand: Hardware and Software Hierarchy. *Journal of Robotic Systems*, 7(5):759–790, 1990.

[117] S.A. Stansfield. Robotic Grasping of Unknown Objects: A Knowledge based Approach. *International Journal of Robotics Research*, 10(4):314–326, 1991.

[118] G.P. Starr. An Experimental Investigation of Object Stiffness Control Using a Multifingered Hand. *Robotics and Autonomous Systems*, 10:33–42, 1992.

[119] T.J. Tarn, A.K. Bejczy, and X. Yun. Design of Dynamic Control of Two Cooperating Robot Arms: Closed Chain Formulation. In *Proceedings 1987 IEEE Conference on Robotics and Automation*, pages 1–7, 1987.

[120] Barrett Technologies. *BarrettHand BH8-250 User's Manual*. Barrett Technologies, Inc., 1997.

[121] M. Teichmann. A Grasp Metric Invariant under Rigid Motions. In *Proceedings 1996 IEEE International Conference on Robotics and Automation*, pages 2143–2148, Minneapolis, MN, 1996.

[122] R. Tomovic, G.A. Bekey, and W.J. Karplus. A Strategy for Grasp Synthesis with Multifingered Robot Hands. In *Proceedings 1987 IEEE International Conference on Robotics and Automation*, pages 83–89, Raleigh, NC, 1987.

[123] J.C. Trinkle. On the Stability and Instantaneous Velocity of Grasped Frictionless Objects. *IEEE Transactions on Robotics and Automation*, 8(5):560–572, 1992.

[124] J.C. Trinkle, J.M. Abel, and R.P. Paul. An investigation of frictionless enveloping grasping in the plane. *International Journal of Robotics Research*, 7(3):33–51, 1988.

[125] J.C. Trinkle, A.O. Farahat, and P.F. Stiller. Second Order Stability Cells of a Frictionless Rigid Body Grasped by Rigid Fingers. In *Proceedings 1994 IEEE International Conference on Robotics and Automation*, pages 2815–2821, San Diego, CA, 1994.

[126] J.C. Trinkle and R.P. Paul. Planning for dexterous manipulation with sliding contacts. *International Journal of Robotics Research*, 9(3):24–48, 1990.

[127] M. Uchiyama and P. Dauchez. A Symmetric Hybrid Position/Force Control Scheme for the Coordination of Two Robots. In *Proceedings 1988 IEEE International Conference on Robotics and Automation*, pages 350–356, Philadelphia, PA, 1988.

[128] M.A. Unseren and A.J.Koivo. Kinematic relations and dynamic modeling for two cooperating manipulators in assembly. In *Proceedings 1987 IEEE Conference on Systems, Man and Cybernetics*, pages 798–802, 1987.

[129] I.D. Walker. Impact Configurations and Measures for Kinematically Redundant and Multiple Armed Robot Systems. *IEEE Transactions on Robotics and Automation*, 10(5):670–683, 1994.

[130] I.D. Walker. A Successful Miltifingered Hand Design - The Case of The Raccoon. In *Proceedings 1995 IEEE/RSJ International Conference on Intelligent Robots and Systems (IROS)*, pages 186–193, Pittsburgh, PA, 1995.

[131] I.D. Walker, R.A. Freeman, and S.I. Marcus. Distribution of Dynamic Loads for Multiple Cooperating Robot Manipulators. *Journal of Robotic Systems*, 6:35–48, 1989.

[132] I.D. Walker, R.A. Freeman, and S.I. Marcus. Analysis of Motion and Internal Loading of Objects Grasped by Multiple Cooperating Manipulators. *International Journal of Robotics Research*, 10(4):396–409, 1991.

[133] J.T. Wen and K. Kreutz. Motion and force control for multiple cooperative manipulators. In *Proceedings 1989 IEEE Conference on Robotics and Automation*, pages 1246–1251, 1989.

[134] G. Wohlke. A Programming and Simulation Environment for the Karlsruhe Dextrous Hand. *Journal of Robotics and Autonomous Systems*, 9:243–263, 1990.

[135] N. Xi, T.J. Tarn, and A.K. Bejczy. Event-Based Planning and Control for Multi-Robot Coordination. In *Proceedings 1993 IEEE International Conference on Robotics and Automation (Vol. 1)*, pages 251–258, Atlanta, GA, 1993.

[136] Y. Xue and I. Kao. Dextrous Sliding Manipulation Using Soft Fingertips. In *Proceedings 1994 IEEE International Conference on Robotics and Automation*, pages 3397–3402, San Diego, CA, 1994.

[137] T. Yoshikawa. Passive and Active Closures by Constraining Mechanisms. In *Proceedings 1996 IEEE International Conference on Robotics and Automation*, pages 1477–1484, Minneapolis, MN, 1996.

[138] T. Yoshikawa and K. Nagai. Manipulating and Grasping Forces in Manipulation by Multifingered Robot Hands. *IEEE Transactions on Robotics and Automation*, 21(1):67–77, 1991.

[139] X. Yun, V. Kumar, N. Sarkar, and E. Paljug. Control of Multiple Arms with Rolling Constraints. In *Proceedings 1992 IEEE International Conference on Robotics and Automation*, pages 2193–2198, Nice, France, 1992.

[140] H. Zhang, K. Tanie, and H. Maekawa. Dextrous Manipulation Planning by Grasp Transformation. In *Proceedings 1996 IEEE International Conference on Robotics and Automation*, pages 3055–3060, Minneapolis, MN, 1996.

[141] X.Y. Zhang, Y. Nakamura, K. Goda, and K. Yoshimoto. Robustness of Power Grasp. In *Proceedings 1994 IEEE International Conference on Robotics and Automation*, pages 2828–2836, San Diego, CA, 1994.

[142] Y. Zhang and W.A. Gruver. Definition and Force Distribution of Power Grasp. In *Proceedings 1995 IEEE International Conference on Robotics and Automation*, pages 1373–1378, Nagoya, Japan, 1995.

[143] Y.F. Zheng and J.Y.S. Luh. Optimal load distribution for two industrial robots handling a single object. In *Proceedings 1988 IEEE Conference on Robotics and Automation*, 1988.

[144] N.B. Zumel and M.A. Erdmann. Nonprehensile Two Palm Manipulation with Non-Equilibrium Transitions between Stable States. In *Proceedings 1996 IEEE International Conference on Robotics and Automation*, pages 3317–3323, Minneapolis, MN, 1996.

Chapter 6

Grasping optimization and control

Grasping, regrasping are difficult operations requiring optimal coordination and control of the fingers. Paper gives a concept and applies it to a four-fingered hand. All fingers are equal and driven by hydraulic actuators. Comparison of theory and measurements are convincing.

6.1 Introduction

Grasping may be looked at as a process of multiple robots, the fingers, being in contact with some object. Therefore, a description of grasping must include the organization of multiple fingers and in addition the contact phenomena. As grasping by an artificial hand is rather slow we shall neglect in this first approach the dynamical aspects and focus on an optimization of grasping strategies and on the control of a hand with four fingers being modeled kinematically and quasi-statically only.

The first step consists in an optimization of the grasp strategy. From trials with five grasp criteria the best one is evaluated. Best performance is achieved by a minimization of the finger force differences with the additional constraints that force and torque equilibrium is maintained, that contact remains established and that the finger forces are within the friction cone. Starting with this basic optimization problem various additional constraints are included: stability of grasping, relative distances between the fingers, sliding of fingers and changing a finger's contact position. The last operation is the most difficult one including some more constraints which express the necessities that the new contact point can be reached, that the

161

fingers cannot penetrate the object and that no finger has a collision with another finger.

In a second step and on the basis of above results another idea is realized which we call hand planning. It optimises the clearance of motion of each finger and the complete finger arrangement, and it regards additional constraints like finger positioning at the object, penetration aspects, the best finger arrangement and the best orientation and location of the grasping plane. With the tools of the two first steps we are able to establish in a third step a typical manipulation planning, grasp planning and hand planning.

All methods are verified experimentally using a hand with hydraulically driven fingers. This fingers have good positioning accuracy and very sensible force control. Maximum speed is about 0.5 sec for a closing/opening process. The size is near a man's finger size. A kind of damping control has been realized based on a oil model, which works without problems.

The first famous artificial hands have been developed in USA and Japan. The UTAH/MIT-Hand [1], the Stanford/JPL-Hand [6] and the WASEDA-Hand are all based on tension-cable-drive-systems, which assure good positioning accuracies and fast motion but not so good force control. In addition cable hands are difficult to design. Up to now direct drives are not small enough with respect to power efficiency, therefore another solution might be a pneumatically or hydraulically driven hand, where hydraulics possesses the advantage of a better density ratio [3]. In the following we shall consider a hydraulic solution.

The hand hardware is one side, the hand software the other one. Grasping, regrasping and manipulation with several fingers require straight and definite strategies which include all physical and geometrical conditions usually connected with processes of that kind. Equilibrium, contact with impacts and friction, questions of reachability, penetration, collision avoidance are some of the essential aspects. In recent years worldwide research focussed on some of these aspects but a comprehensive solution is still missing and, as a matter of fact, still far away of the perfect behavior of the human hand. Strategies of the kind must not only calculate the finger forces necessary to manipulate the object [5], but also locate the fingers on the object in such a way that a stable grasp can be achieved [4]. With a few exceptions [2], the work on grasp planning has focused on one aspect or the other. In this paper, a grasp strategy is demonstrated which accomplishes both tasks. Given the desired external forces on the object and the object geometry, the strategy calculates the grasp points and the finger forces necessary to achieve the desired external wrench on the object.

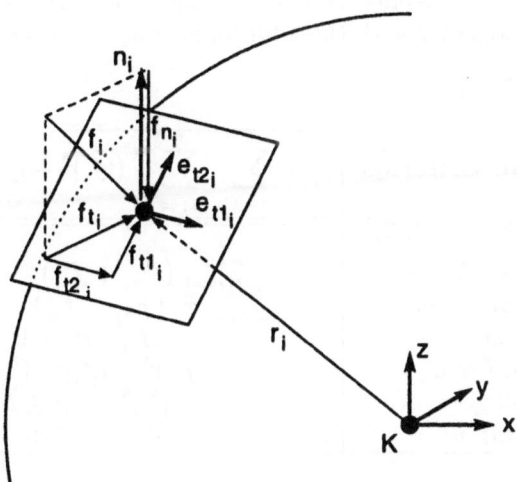

Figure 6.1: Decomposition of finger forces.

6.2 Grasp strategies

Finger forces have been decomposed in a first step into components which are normal and tangential to the plane of contact. This deviates from the decomposition into manipulation and internal forces [8], but is more convenient for mechanical reasons. According to Figure 6.1 we then write

$$\boldsymbol{f}_{n_i} = f_{n_i}\boldsymbol{n}_i, \qquad \boldsymbol{f}_{t_i} = f_{t1_i}\boldsymbol{e}_{t1_i} + f_{t2_i}\boldsymbol{e}_{t2_i}, \qquad \boldsymbol{f}_i = \boldsymbol{f}_{n_i} + \boldsymbol{f}_{t_i} \qquad (6.1)$$

The second problem involves an optimization criterion for an evaluation of the finger forces. Five criteria have been investigated [7]: minimum dependence on the friction coefficients, minimum tangential finger forces, minimal sum of all finger force magnitudes, minimum of the maximal finger force, minimum difference of the finger force magnitudes. It turns out that the last criterion gives a best approach for a good distribution of the forces over all fingers. Therefore, for all further considerations finger forces are optimally selected according to the criterion

$$G = \sum_{i=1}^{n} \sum_{\substack{j=1 \\ (j \neq 1)}}^{n} \left(|\boldsymbol{f}_i|^2 - |\boldsymbol{f}_j|^2 \right)^2 \implies \min! \qquad (6.2)$$

Three different optimization processes are considered, normal grasping with stability margins and sufficient finger distances, grasping with controlled sliding and grasping with regrasping. The corresponding optimization processes together with the additional constraints are the following:

- Normal Grasping

Optimization Criterion	$G = \sum\limits_{i=1}^{n} \sum\limits_{j=1(j\neq 1)}^{n} \left(|f_i|^2 - |f_j|^2\right)^2 \longrightarrow \min$
Necessary Conditions Force Equilibrium Moment Equilibrium Contact Friction Cone Stability Separation	$\sum_{i=1}^{n}\left(f_{n_i} + f_{t_i}\right) - F_e = 0$ $\sum_{i=1}^{n} \tilde{r}_i\left(f_{n_i} + f_{t_i}\right) - M_e = 0$ $f_{n_i} \cdot n_i < 0$ $|f_{t_i}|^2 - \mu^2|f_{n_i}|^2 < 0$ $|\sum_{i=n}^{n} n_i| \leq S$ $|r_i - r_j| - \epsilon_{\min} \geq 0 \qquad i \neq j$

- Grasping with Controlled Sliding (see Figure 6.2)

Optimization Criterion	$G = \sum\limits_{i=1}^{n} \sum\limits_{j=1(j\neq 1)}^{n} \left(|f_i|^2 - |f_j|^2\right)^2 \longrightarrow \min$
Necessary Conditions Force Equilibrium Moment Equilibrium Contact Friction Cone Sliding Direction Sliding Forces	$\sum_{i=1}^{n}\left(f_{n_i} + f_{t_i}\right) - F_e = 0$ $\sum_{i=1}^{n} \tilde{r}_i\left(f_{n_i} + f_{t_i}\right) - M_e = 0$ $f_{n_i} \cdot n_i < 0$ $|f_{t_i}|^2 - \mu^2|f_{n_i}|^2 < 0$ $d = d_{t1}e_{t1} + d_{t2}e_{t2}$ $f_{nr} = -k_r/\mu \quad \text{with} \quad k_r \geq 0$ $f_{t1_r} = k_r d_{t1}$ $f_{t2_r} = k_r d_{t2}$

- Grasping with Regrasping

Optimization Criterion	$G = \sum\limits_{i=1}^{n} \sum\limits_{j=1(j\neq 1)}^{n} \left(|f_i|^2 - |f_j|^2\right)^2 \longrightarrow \min$
Necessary Conditions Force Equilibrium Moment Equilibrium Contact Friction Cone Regrasping	$\sum_{i=1}^{n}\left(f_{n_i} + f_{t_i}\right) - F_e = 0$ $\sum_{i=1}^{n} \tilde{r}_i\left(f_{n_i} + f_{t_i}\right) - M_e = 0$ $f_{n_i} \cdot n_i < 0$ $|f_{t_i}|^2 - \mu^2|f_{n_i}|^2 < 0$ • Reachability • No Penetration • No Collision

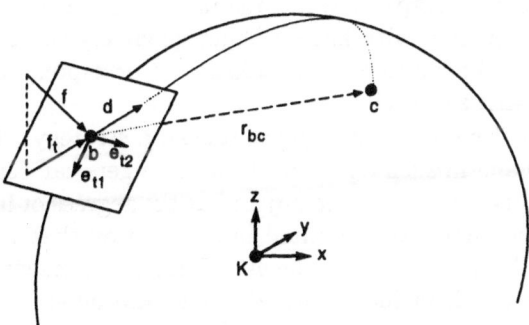

Figure 6.2: Grasping with sliding from b to c.

The meaning of the various conditions is evident. Neglecting inertia forces the finger forces and the external forces due to gravity must be in static equilibrium. The same is true for the torques ($\tilde{a}b = a \times b$ definition of cross product). The contact condition says that the finger forces normal to the contact plane must be negative to assure always pressure forces only. Furtheron the finger forces must be within the friction cone to avoid uncontrolled sliding.

The normal vectors to the object's surface at the grasp points provide a good insight into the stability of the grip: the smaller the sum of the vectors, the more stable the grasp. The grasp is less stable in the direction opposite the resulting sum, which means that it is less capable of resisting disturbances in that direction. This stability writes

$$|\sum_{i=1}^{n} n_i| \leq S \ , \tag{6.3}$$

where S is the desired stability measure.

The separation condition guarantees that a minimum separation is maintained between the grasp points, so that the fingers do not come too close to one another. For grasping with controlled sliding the sliding direction is given by a direct connection to the target point (point c in Figure 6.2). The sliding forces follow the geometry and are controlled by a constant magnitude $k_r \geq 0$.

For regrasping questions of reachability, penetration and collision become important. Normal grasping and grasping with sliding can be performed with three fingers, for regrasping we need at least four fingers. Given the object and the geometry of the fingers we decide geometrically with the help of the fingers' workspaces what points can be reached without violating stability. Furtheron, with known finger geometry we also can evaluate the two problems of penetration and collision. Corresponding formulas and methods are described in [7].

In order to automate the grasping process, a strategy which can orient and locate the hand in such a manner that all fingers can reach their designated grasp points is needed. The object has six degrees of freedom relative to the hand which have to be limited in such a way that the grasp points are reachable. To solve these problems of hand placement a method has been developed which includes several steps: the definition of the grasp-triangle, a rough hand orientation, the finger assignment, and, finally, an optimization of the hand orientation and distance to the object.

Before evaluating these data the following geometric quantities must be known:

- Hand Geometry
  (position and orientation of the fingers on the palm described in hand frames)

- Workspace
  (position and orientation of the robot base described in a robot coordinate frame)

- Path planning
  (position and orientation of the object in a tool frame)

- Grasp Points
  (position of the i-th grasp point in a body-fixed object frame)

- Hand Orientation
  (position and orientation of the robot hand)

With these data known one must check in a first step by applying inverse finger kinematics if the grasp point can be reached without penetrating the object. In a second step position and orientation of the hand are calculated by arranging the palm surface parallel to the grasp triangle and the palm center over the grasp center. Then in a third step the orientation and the distance of the hand are optimized by maximizing the remaining workspace of the fingers.

The last step consists in a planning procedure for a manipulation process which includes all sequences of path planning, grasp planning and hand planning. Figure 6.3 indicates the corresponding strategy [7].

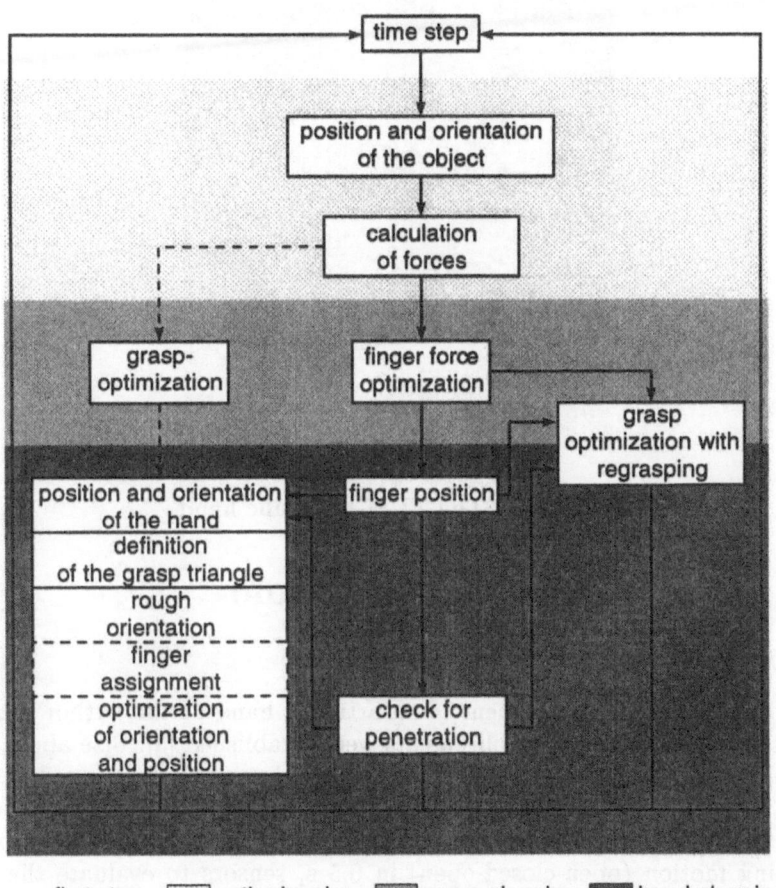

Figure 6.3: Manipulation planning.

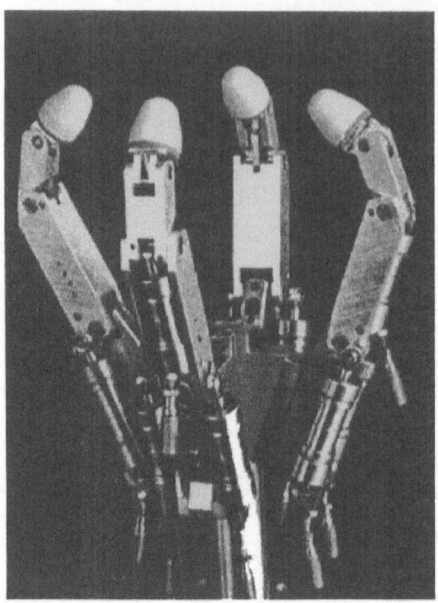

Figure 6.4: The TUM-hydraulic hand.

6.3 The TUM-hydraulic hand

6.3.1 The design

When starting the development of an artificial hand at the author's institute the following design requirements were established [3]: Size about the human hand, three to four equal fingers which can be exchanged easily, three degrees of freedom per finger, maximum manipulation weight at least 10 N and minimum about 1 N, individual finger force 30 N, one complete grasping motion (open-closed-open) in 0.5 s, sensors to evaluate the fingertip forces with respect to amount, direction and location. A trade-off study with various drive systems (pneumatic, hydraulic, electric, cables) results in a solution with hydraulic drives. They allow excellent force control in a wide range of force magnitudes, on the other hand they have some disadvantages like leakage and difficult calibration. Figure 6.4 gives an impression of a four-finger arrangement, and Figure 6.5 shows one finger in more detail [3,7]. The fingers are fixed to the palm by two screws only which allows a quick change of the finger-palm-combination.

All fingers are equal, and each one possesses three degrees of freedom,

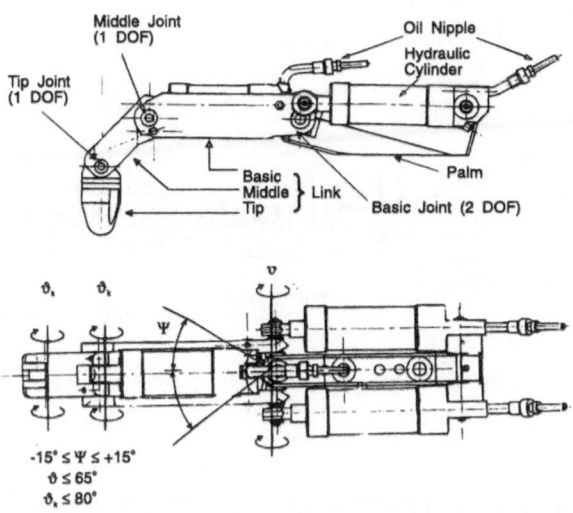

Figure 6.5: Design of the hydraulic finger [3].

one combined degree of freedom for the first two finger joints and additional two degrees of freedom at the finger's root. From this we have realized two DOF in the finger plane and one DOF to allow a motion of the finger plane itself (Figure 6.5).

The fingers are driven by hydraulic cylinders which operate in one direction by oil pressure and in the opposite direction by a prestressed spring. The tip and middle links are connected by a simple mechanism combining them to one DOF. The basic joint is driven by two cylinders which can generate two DOF. Altogether this results in three degrees of freedom $\varphi_1, \varphi_2, \varphi_3$. The finger arrangement of Figure 6.5 has a size like a middle finger of a human hand.

6.3.2 Measurement and control

Measurement and control of the hydraulic finger is realized in the following way, which again represents the outcome of an investigation concerning a large variety of possible solutions.

The piston is driven by oil pressure on one side and by a prestressed spring on the opposite side (Figure 6.6). The oil is moved through a 4 m long elastic tube from the hydraulic power station to the piston. The hydraulic

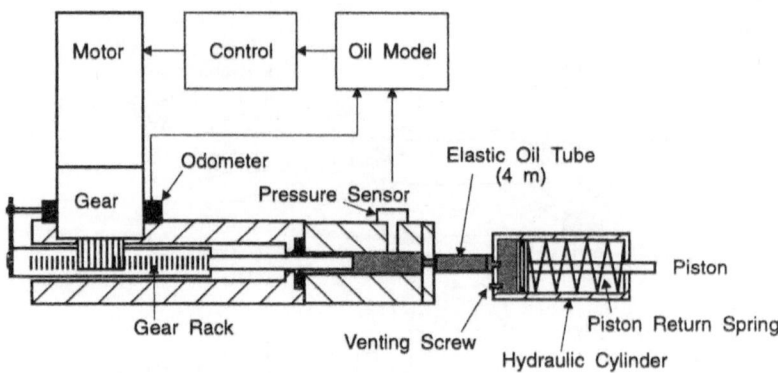

Figure 6.6: The hydraulic finger control [3].

power station consists of a motor-gear-combination which drives a gear rack with a piston. This piston moves the oil within a cylinder and from there to the elastic oil tube.

Two measurements are installed. Firstly, an odometer measures the location of the gear rack and with it of the oil piston, which gives an information about the position of the oil column in the cylinder-tube-cylinder combination. Secondly, a pressure sensor measures the oil pressure at the exit of the driving cylinder to the tube. Direct measurements at the finger cylinders are not implemented due to the requirement of having only one connection for each finger cylinder to the ground supported power station.

With these two measurements the motor in Figure 6.6 cannot be controlled. We need in addition an oil model which takes into account all pressure losses and friction forces from the power station to the finger cylinders. Such a model is used as indicated in Figure 6.6, therefore it should be as simple as possible. Figure 6.7 depicts the principal modeling which represents a typical situation for cyclic motion.

Increasing the pressure by moving the gear rack we walk along characteristic 1. When the pressure time derivative $\dot{p}$ changes sign then the finger piston sticks and its position x_F and its piston force F_K remain constant (characteristic 2). This state is maintained until all external forces like oil pressure force, piston force, spring force are large enough to overcome the stiction state and then to drive the finger piston in the opposite direction. The pressure decreases along the characteristic 3. The piston again sticks when $\dot{p}$ will change sign and x_F, F_K will be constant along characteristic

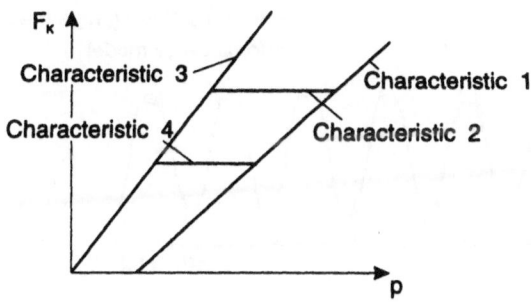

Figure 6.7: Oil model.

4. The two characteristics 1, 3 follow the simple equations

$$F_K = k_1 x_A + k_2 p + F_r sgn(\dot{x}_F),$$
$$x_F = k_3 x_A + k_{4_{1,3}} p, \qquad \text{with } F_r = F_{r_0} + c_r p \qquad (6.4)$$

where the coefficients are partly determined by experiments [3]. The sign of $\dot{x}_F$ is given with the angular speed of the motor. The four switching points in Figure 6.7 can also be evaluated by considering sign $(\dot{x}_F)$. If the velocity $\dot{x}_F$ changes sign, the pressure derivative $\dot{p}$ will change sign as well, at least for the relative slow motion as considered in this case.

For a verification of this oil model we press the finger piston against a bending bar with a strain gauge arrangement. We compare these measurements with the forces recalculated from the oil model. Figure 6.8 gives a comparison for position x_F and force F_K.

The advantages of the solution are obvious. The basic drive is the configuration of Figure 6.6, which is the same for all fingers. Each finger possesses three hydraulic drives of that type, and each hand might have any number of equal fingers. The number of connections of the fingers and the ground station is minimized, and all drives are rather simple. Nevertheless any complicated grasping program might be executed by these fingers [3,7].

To execute a complete grasping program we need a supervisory control of each finger cooperating together and performing the grasping sequences, and we need a planning process for manipulating an object with the fingers. Without going into details [3,7], we present two schemes. The first one of Figure 6.9 illustrates the hardware of the TUM-hand. All four fingers and all drives of the fingers are connected by a VME-Bus-System which combines a SUN-workstation, a 486 CPU-PC-computer and several AD- and DA-converters. The converters receive the measurement signals and

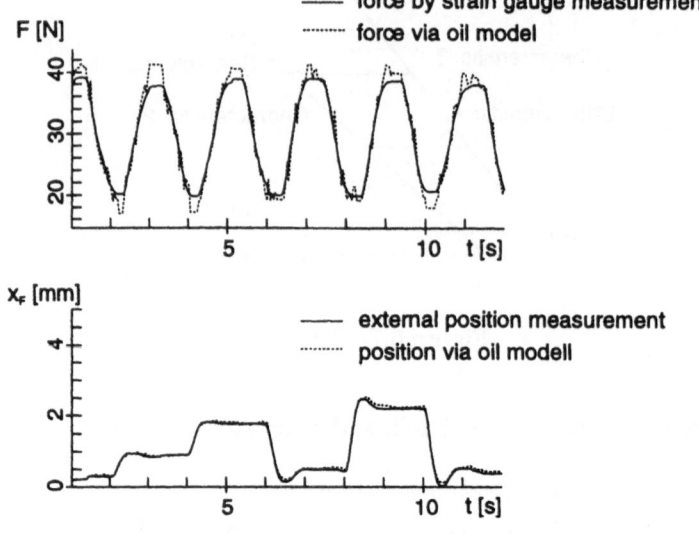

Figure 6.8: Verification of the oil model.

send signals to the finger drives. This set-up allows control of the complete hand.

6.4 Examples

On the basis of the optimizations in the grasping chapter and of the planning procedures (Figure 6.3) several simulations have been performed to show the efficiency of the methods in grasping and regrasping [7]. As one typical example we show here the rotation of a sphere by regrasping with a four-fingered hand. A typical grasp pattern as developed in [7] is given with Figure 6.10, which is self-explaining. The sequence of finger positions in performing this task is illustrated by the pictures of Figure 6.11. We see that the above discussed optimizations generate meaningful sequences of finger operations.

The theories for grasping and for the hand, the finger design and the hand-hardware are verified by experiments, rotation of an ellipsoid, regrasping of a cuboid and manipulation of a raw egg. The last mentioned experiment also has been presented at the Hannover Industrial Fair 1994. We show here only the regrasping experiment for a cuboid which is held against gravity. Its weight amounts to 195 g, its size is 15 × 25 × 40 mm.

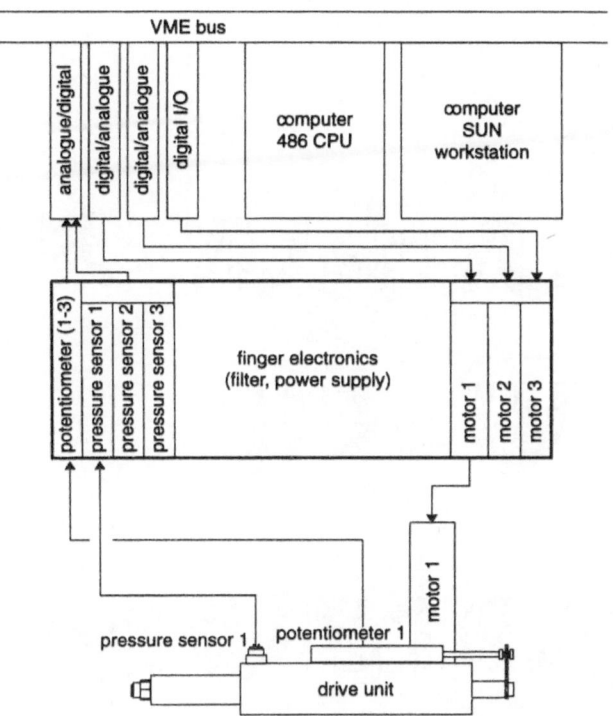

Figure 6.9: Hardware scheme of the TUM-hydraulic hand.

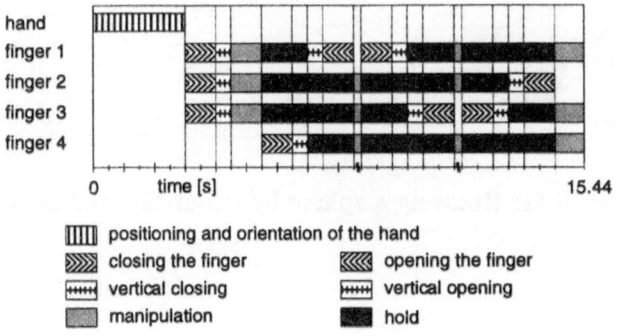

Figure 6.10: Grasping pattern [7].

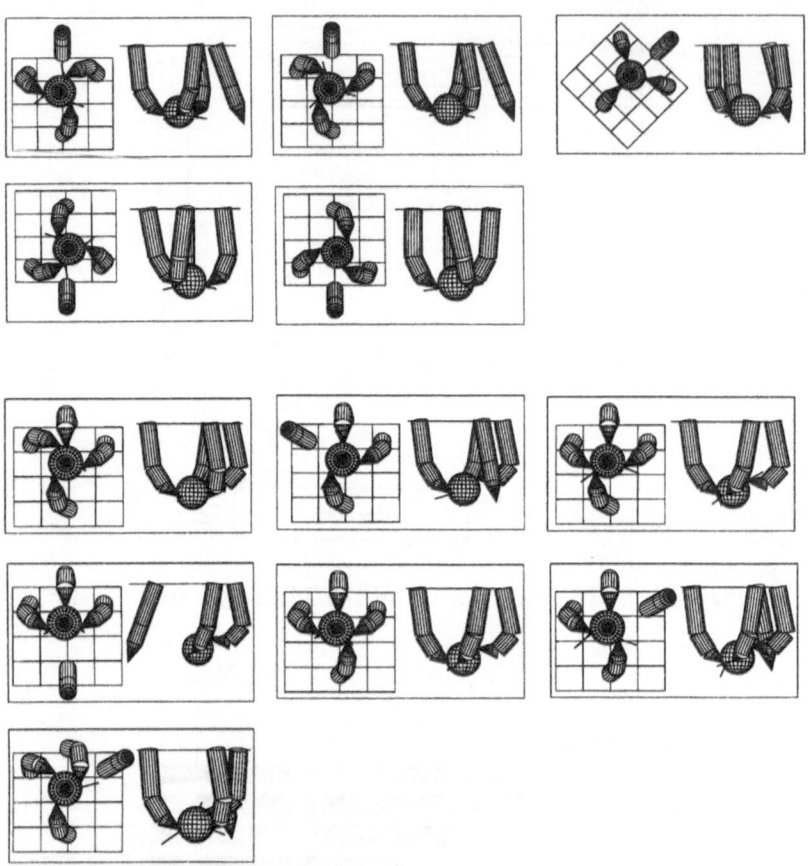

Figure 6.11: Rotating a sphere by a four-fingered hand [7].

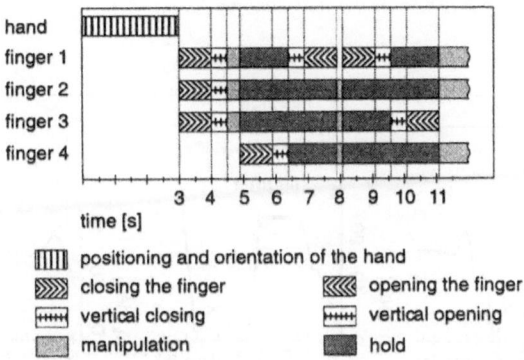

Figure 6.12: Grasp pattern for regrasping a cuboid.

The static friction coefficient amounts to $\mu = 0.4$. Two regrasping steps are performed (see pattern, Figure 6.12).

Figure 6.13 depicts a comparison of theoretical finger planning according to Figure 6.12 with experimental measurements of the finger angles (Figure 6.6). Theory and experiments go very well together.

6.5 Conclusions

Strategies for cooperating fingers of an artificial hand are considered. Finger force adaptation is carried through by minimizing the finger force differences between the fingers and by taking into account certain constraints like equilibrium, friction, contact, stability, sliding motion, reachability, penetration, collision. A manipulation planning considers path planning of an object, grasp planning of the fingers and hand positioning planning.

Grasp experiments are performed with a hydraulically driven hand with three and four fingers for which design and control concept are verified experimentally. Each finger possesses three degrees of freedom which are controlled by a ground station. It is connected by a 4 m long oil tube with the finger. All experiments agree well with theory.

Acknowledgments

Funding for this work was provided by the German Ministry for Research and Technology (BMFT) through the SEKON Project (Grant: 01 IN 104 D/7).

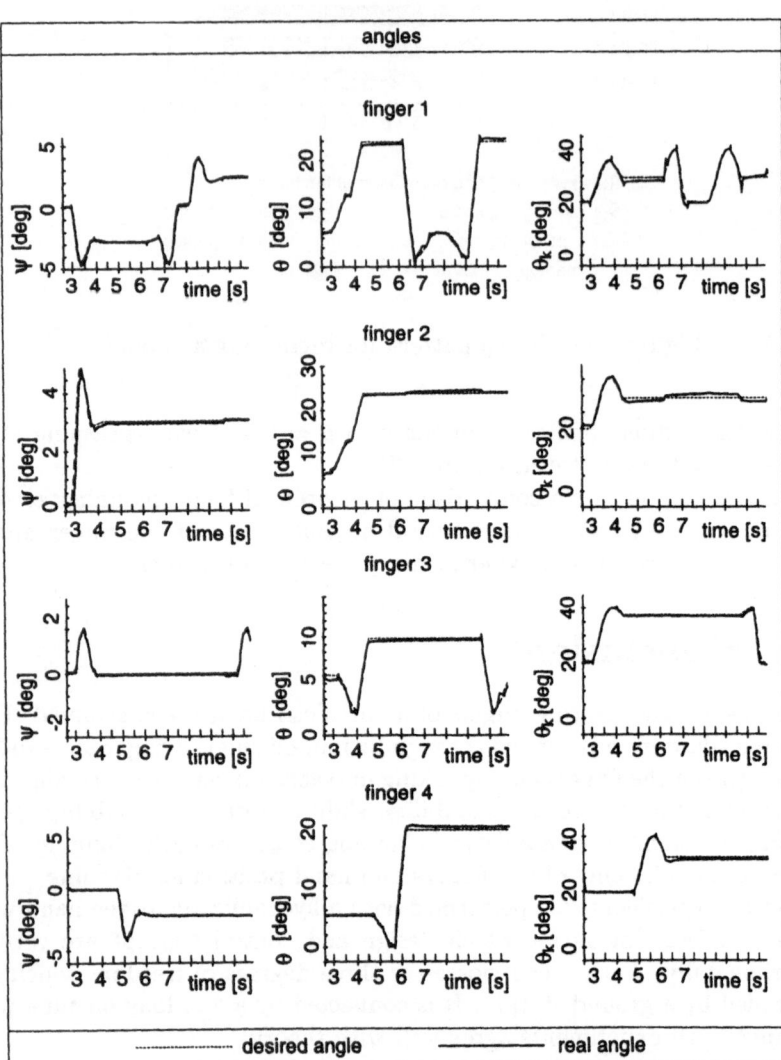

Figure 6.13: Theory and measurements for regrasping a cuboid, angular motion of the fingers.

References

[1] Biggers, K.B., Jacobsen, S.C. and Gerpheide, G.E.: Low Level Control of the Utah/MIT Dextrous Hand. *Proc. of IEEE Int. Conf. on Robotics and Automation* (April 1987).

[2] Li, Z., Hsu, P. and Sastry, S.S.: Grasping and Coordinated Manipulation by a Multifingered Robot Hand. *Int. Journal of Robotics Research*, 8(4)(1989), 33-50.

[3] Menzel, R.: Konstruktion und Regelung einer hydraulischen Hand. *Fortschrittberichte VDI*, Reihe 8, Nr. 451, VDI-Verlag Düsseldorf (1995).

[4] Omata, T.: Fingertip position of a multifingered Hand. *Proc. of IEEE Int. Conf. on Robotics and Automation*, (May 1990).

[5] Park, Y.C.: Grasping and Manipulation of an Object using a Multifingered Robot Hand. *PhD thesis*, University of New Mexico, (May 1990).

[6] Salisbury, J.K.: Kinematic and Force Analysis of Articulated Hands. *PhD thesis*, Stanford University, (May 1982).

[7] Woelfl, K.: Planung von Manipulationsvorgängen einer Roboterhand. *Fortschrittberichte VDI*, Reihe 8, Nr. 455, VDI-Verlag Düsseldorf (1995).

[8] Yoshikawa, T. and Nagai, K.: Manipulating and Grasping Forces in Manipulation by Multi-Fingered Hands. *Proc. of Int. Conference on Robotics and Automation*, pp. 1998-2004, Raleigh, (March 1987).

Lecture Notes in Control and Information Sciences

Edited by M. Thoma

Vol. 220: Brogliato, B.
Nonsmooth Impact Mechanics: Models,
Dynamics and Control
420 pp. 1996 [3-540-76079-2]

Vol. 221: Kelkar, A.; Joshi, S.
Control of Nonlinear Multibody Flexible
Space Structures
160 pp. 1996 [3-540-76093-8]

Vol. 222: Morse, A.S.
Control Using Logic-Based Switching
288 pp. 1997 [3-540-76097-0]

Vol. 223: Khatib, O.; Salisbury, J.K.
Experimental Robotics IV: The 4th
International Symposium, Stanford,
California,
June 30 - July 2, 1995
596 pp. 1997 [3-540-76133-0]

Vol. 224: Magni, J.-F.; Bennani, S.;
Terlouw, J. (Eds)
Robust Flight Control: A Design Challenge
664 pp. 1997 [3-540-76151-9]

Vol. 225: Poznyak, A.S.; Najim, K.
Learning Automata and Stochastic
Optimization
219 pp. 1997 [3-540-76154-3]

Vol. 226: Cooperman, G.; Michler, G.;
Vinck, H. (Eds)
Workshop on High Performance Computing
and Gigabit Local Area Networks
248 pp. 1997 [3-540-76169-1]

Vol. 227: Tarbouriech, S.; Garcia, G. (Eds)
Control of Uncertain Systems with Bounded
Inputs
203 pp. 1997 [3-540-76183-7]

Vol. 228: Dugard, L.; Verriest, E.I. (Eds)
Stability and Control of Time-delay Systems
344 pp. 1998 [3-540-76193-4]

Vol. 229: Laumond, J.-P. (Ed.)
Robot Motion Planning and Control
360 pp. 1998 [3-540-76219-1]

Vol. 230: Siciliano, B.; Valavanis, K.P. (Eds)
Control Problems in Robotics and
Automation
328 pp. 1998 [3-540-76220-5]

Vol. 231: Emel'yanov, S.V.; Burovoi, I.A.;
Levada, F.Yu.
Control of Indefinite Nonlinear Dynamic
Systems
196 pp. 1998 [3-540-76245-0]

Vol. 232: Casals A.; de Almeida, A.T.
Experimental Robotics V: The Fifth
International Symposium Barcelona,
Catalonia, June 15-18, 1997
190 pp. 1998 [3-540-76218-3]